# *IN VITRO* HAPLOID PRODUCTION IN HIGHER PLANTS

Current Plant Science and Biotechnology in Agriculture

VOLUME 26



*Aims and Scope*
The book series is intended for readers ranging from advanced students to senior research scientists and corporate directors interested in acquiring in-depth, state-of-the-art knowledge about research findings and techniques related to all aspects of agricultural biotechnology. Although the previous volumes in the series dealt with plant science and biotechnology, the aim is now to also include volumes dealing with animals science, food science and microbiology. While the subject matter will relate more particularly to agricultural applications, timely topics in basic science and biotechnology will also be explored. Some volumes will report progress in rapidly advancing disciplines through proceedings of symposia and workshops while others will detail fundamental information of an enduring nature that will be referenced repeatedly.

*The titles published in this series are listed at the end of this volume.*

# *In Vitro* Haploid Production in Higher Plants

*Volume 4 – Cereals*

*Edited by*

S. MOHAN JAIN

*Plant Production Department, University of Helsinki, Helsinki, Finland*

S.K. SOPORY

*School of Life Science, Jawaharlal Nehru University, New Delhi, India*

and

R.E. VEILLEUX

*Department of Horticulture, Virginia Polytechnic Institute and State University, Blacksburg, Virginia, U.S.A.*

SPRINGER-SCIENCE+BUSINESS MEDIA, B.V.

**Library of Congress Cataloging-in-Publication Data**

In vitro haploid production in higher plants / editors, S. Mohan Jain, S.K. Sopory, R.E. Veilleux.
p. cm. -- (Current plant science and biotechnology in agriculture ; v. 23)
Includes index.
Contents: v. 1. Fundamental aspects
ISBN 978-90-481-4682-6 ISBN 978-94-017-1862-2 (eBook)
DOI 10.1007/978-94-017-1862-2
(hardback : alk. paper)
1. Micropropagation. 2. Haploidy. 3. Crops--Genetic engineering. 4. Plant breeding. I. Jain, S. Mohan. II. Sopory, S. K. III. Veilleux, R. E. IV. Series: Current plant science and biotechnology in agriculture ; 23.
SB123.6.I45 1996
631.5'23--dc20 95-304

ISBN 978-90-481-4682-6

*Printed on acid-free paper*

# Table of Contents

# Haploid production in higher plants
*A dedication*

INDRA K. VASIL

The value of haploids in genetic analysis and plant breeding has been known for a long time. Natural haploid embryos and plants, derived from gametophytic cells, have been described in about 100 species of angiosperms. However, haploids occur only rarely in nature. To be useful, they must be produced in large numbers. Therefore, many attempts have been made over the years to increase the efficiency of *in ovulo* haploid production, but none of these has proven to be of wide practical utility. The early attempts to obtain haploid plants from the male gametophyte of gymnosperms (Tulecke, 1953) and angiosperms (Yamada *et al.*, 1963) resulted only in the production of haploid callus tissues (Vasil, 1980).

Embryo-like structures formed in cultured anthers of *Datura innoxia* were first described by Guha and Maheshwari (1964). They were considered to have originated from the somatic tissues of the anther. In a subsequent study, it was determined that the somatic embryos and the resulting plantlets were indeed derived from the developing microspores and were haploid in nature (Guha and Maheshwari, 1966). As is true of most pioneering studies, these first androgenic haploids were neither grown to maturity, nor were the experimental conditions for their production clearly defined. Their real value was in demonstrating the feasibility of the experimental production of haploids. Haploid plants were soon obtained from cultured anthers of *Nicotiana sylvestris* and *N. tabacum* by Bourgin and Nitsch (1967). These and subsequent studies by Nitsch and Nitsch (1969) clearly established that the culture of excised anthers at a precise stage of development was the most important requirement for switching the development of pollen from a gametophytic to a sporophytic phase, resulting in the formation of haploid embryos and/or plants. They also described a simple nutrient medium for the culture of anthers, and an easy procedure for obtaining dihaploid homozygous plants. The elegant, simple and reliable method of haploid production invented by Jean Pierre Nitsch and his associates provided much stimulus for future studies by many others.

During the past three decades many improved methods as well as nutrient media have been developed to increase the efficiency of production of andro-

Prof. S.C. Maheshwari

Dr. J.P. Nitsch

genic haploids, from cultured anthers as well as isolated microspores, in a wide variety of species. Success has also been achieved in obtaining gynogenic haploids from cultured ovaries or ovules. As a result, haploids are being used increasingly and profitably in breeding programmes for the development of new and improved cultivars. The various chapters in this and the companion volumes describe in detail the basic as well as many applied aspects of haploid production and utilization.

It has been my pleasure and privilege to have known the late Jean Pierre Bourgin, Sipra Guha-Mukherjee, Satish C. Maheshwari, Colette Nitsch and the late Jean Pierre Nitsch, all pioneers in haploid research. These volumes are dedicated to them for their seminal contributions to the experimental production of haploids and for creating a whole new field of basic and applied plant research.

## References

Bourgin, J.P. and J.P. Nitsch, 1967. Obtention de *Nicotiana* haploïdes à partir d'étamines cultivées *in vitro*. Ann. Physiol. Veg. 9: 377–382.

Guha, S. and S.C. Maheshwari, 1964. *In vitro* production of embryos from anthers of *Datura*. Nature 204: 497.

Guha, S. and S.C. Maheshwari, 1966. Cell division and differentiation of embryos in the pollen grains of *Datura in vitro*. Nature 212: 97–98.

Nitsch, J.P. and C. Nitsch, 1969. Haploid plants from pollen grains. Science 163: 85--87.

Tulecke W., 1953. A tissue derived from the pollen of *Ginkgo biloba*. Science 117: 599–600.

Yamada, T., T. Shoji and Y. Sinoto. 1963. Formation of calli and free cells in the tissue culture of *Tradescantia reflexa*. Bot. Mag. Tokyo 76: 332–339.

Vasil, I.K., 1980. Androgenetic haploids. Int. Rev. Cytol. Suppl. 11A: 195–223.

# General Preface

Since the beginning of agricultural production, there has been a continuous effort to grow more and better quality food to feed ever increasing populations. Both improved cultural practices and improved crop plants have allowed us to divert more human resources to non-agricultural activities while still increasing agricultural production. Malthusian population predictions continue to alarm agricultural researchers, especially plant breeders, to seek new technologies that will continue to allow us to produce more and better food by fewer people on less land. Both improvement of existing cultivars and development of new high-yielding cultivars are common goals for breeders of all crops. *In vitro* haploid production is among the new technologies that show great promise toward the goal of increasing crop yields by making similar germplasm available for many crops that was used to implement one of the greatest plant breeding success stories of this century, i.e., the development of hybrid maize by crosses of inbred lines. One of the main applications of anther culture has been to produce diploid homozygous pure lines in a single generation, thus saving many generations of backcrossing to reach homozygosity by traditional means or in crops where self-pollination is not possible.

Because doubled haploids are equivalent to inbred lines, their value has been appreciated by plant breeders for decades. The search for natural haploids and methods to induce them has been ongoing since the beginning of the 20th century. Blakeslee (1921) first identified naturally occurring haploids of *Datura stramonium* and subsequently, natural haploids of many other plants were reported by various researchers. However, naturally occurring haploids could not be produced in sufficient numbers by reliable techniques for their extensive use in breeding programmes. In 1964, the research group headed by Prof. S.C. Maheshwari, Department of Botany, Delhi University, India, reported haploid production in *Datura innoxia* for the first time by anther culture. Since this discovery, many of the limitations of the technique have been overcome such that it is currently employed for the production of haploids and doubled haploids of many crop species throughout the world. The early contributions of Drs. C. Nitsch and J.-P. Nitsch (France), G. Melchers (Germany), M.S. Swaminathan (India), I.K. Vasil (USA), N.

Sunderland (UK), and Hu Han (China) towards overcoming these limitations and adapting the technology to a variety of crops must be acknowledged. Their realization of the potential of anther culture and tireless pursuit of reliable techniques that would facilitate its success has led to its implementation in breeding programmes.

In addition to its practical applications, *in vitro* haploid extraction has changed our understanding of developmental processes in plants. Androgenesis can be thought of as a type of somatic embryogenesis that involves cells, i.e., microspores, that at first thought would not have been expected to be embryogenically competent. How can the natural course of microsporogenesis be diverted onto an embryogenic pathway? Why are some microspores competent for androgenic development and not others? How can the process of anther culture be a heritable trait in crosses between competent and non-competent parents? Does the process of anther culture impose some selection pressure on the population of microspores or otherwise result in some undesirable change expressed in the population of regenerated plants? Why are albinos so common among the anther-derived regenerants of some species when it is obvious that microspores must contain proplastids in order for green plants to be regenerated at all? We have only begun to answer some of these questions.

This book project was submitted with the consent of co-editors to the Kluwer Academic Publishers, Dordrecht, The Netherlands. The publisher had this project reviewed by anonymous reviewers. Finally, on the basis of the positive comments of the reviewers, the publisher gave us the contract to proceed with this book project. We have not followed any conservative format of chapters and gave all the liberty to the authors to write the way they felt appropriate. Most of the chapters are reviews of work done. However, in some cases, where a lot of work has not been done in the past, the authors have been encouraged to give their own research findings in details.

In this set of volumes, we have made an attempt to assimilate detailed descriptions of various aspects of anther culture and related *in vitro* procedures. Many chapters have been written by experts in the various applications of anther culture to specific crops. In addition to crop-by-crop discussions on the progress of anther culture, we have also included chapters on other topics concerning the utilization of *in vitro* haploids in plants. Embryogenic microspores have recently been regarded as ideal targets for genetic transformation. Molecular markers such as RFLPs, RAPDs, or SSRs can be used to determine disturbed segregation ratios in haploid populations or to tag traits of interest to plant breeders. The potential of pollen protoplasts is discussed. *In vitro* selection during androgenesis, both imposed and inadvertent, is also considered.

The series is divided into five volumes. Volume 1 contains 18 chapters and primarily covers fundamental aspects of haploidy and various methods of haploid extraction, e.g., anther culture, microspore culture, ovary culture, etc. The second volume comprises 21 chapters and describes applications of haploid breeding in protoplast manipulations, mutation breeding, RFLP mapping, identification of quantitative trait loci (QTLs), cryopreservation, chromosome engineering by anther culture, molecular biology of pollen rejection, transformation of pollen/microspores, etc. The third volume has 20 chapters focussed on haploid breeding in selected important crops including vegetables (*Allium* spp., *Brassica* spp., *Capsicum*, *Cichorium*, *Cucumis*, *Solanum melongena*, *Solanum tuberosum*); fruit crops (*Malus*, *Fragaria*, *Vitis*); and other miscellaneous crops (*Beta*, *Coffea*, *Ginkgo*, *Glycine*, *Medicago*, *Saccharum*, *Sinocalamus latiflora*). We have included 11 chapters in the fourth volume on haploid breeding in cereals (wheat, rice, barley, oats, sorghum, maize, triticale, rye, pearl millet, buckwheat). The fifth volume has 13 chapters, mainly dealing with ornamentals, tobacco, tomato, cotton, linseed, sunflower, asparagus, niger, gynogenic haploids in angiosperms and haploids in potato interspecific somatic hybrids.

While preparing these volumes, we were overwhelmed by the enthusiastic response and timely cooperation of invited authors and the many research scientists who gave freely of their time to review the manuscripts. The reviewers were: R.I.S. Brettel (Australia); B.S. Ahloowalia (Austria); J.M. Bonga, K.N. Kao, K. Kasha, L.K. Kott, K.P. Pauls, R. Sadashivaiah (Canada); Hu Han (China); H. Ahokas, V. Kauppinen, J. Peltonen, S. Sarvori, L. Simola, P.M.A. Tigerstedt (Finland); C. Dóre, R.S. Sangwan (France); B. Foroughi-Wehr, D. Hess, S. Deimling, H. Uhrig, G. Wenzel (Germany); S.S. Bhojwani, P.B. Kirti, S.C. Maheshwari, A.F. Mascarenhas, P.S. Nadgauda, D. Pental, S.K. Raina, P.S. Rao, N. Sarin, G. Lakhshmi Sita (India); A. Mujeeb-Kazi (Mexico); H. Dons, K. Sree Ramulu (The Netherlands); D.S. Brar, G.S. Khush (Philippines); M. Zenkteler (Poland); A.M. Vieitez (Spain); C. Bornman, K. Glimelius, A. Wallin (Sweden); W. Chang (Taiwan); J. Dunwell, V.E. Franklin-Tong, W. Powell (UK); P.S. Baenziger, E. Earle, G.J. Galletta, D.J. Gray, P.K. Gupta, J.P. Helgeson, J. Janick, S.M. Reed, H.S. Rines, G. Schaeffer, T.L. Sims, I.K. Vasil, J.M. Widholm (USA).

It is still too early to write the last chapter on *in vitro* haploids. There are those who argue that its great potential will result in improved cultivars of many of our major crops. On the other hand, there are those who think its potential has been overrated, that the severe inbreeding depression observed among primary doubled haploids and the lack of selection pressure for functional sexual flower parts during the process of androgenesis will result only in useless, fruitless plants. The first cultivars employing anther-derived doubled haploids in their pedigree have already been released for a number of crops including wheat, rice, maize, and asparagus. Whether these cultivars and future such releases will endure remains to be seen.

S. Mohan Jain
S.K. Sopory
R.E. Veilleux

# Preface to Volume 4

This fourth volume on *In Vitro* Haploid Production in Higher Plants comprises 11 chapters on cereal haploid production. All chapters are crop specific and each chapter contains an introduction about the selected plant, the techniques (anther culture, microspore culture, ovary/ovule culture) that have been successfully used for haploid production, the factors that have influenced the success of these techniques, the identification and genetic characterization of haploid regenerants, the application of haploids in breeding, and a brief conclusion on the potential of haploid breeding in the specific crop. The chapters review haploidy in cereal crops including rice (*Oryza sativa*), maize (*Zea mays*), wheat (*Triticum aestivum*), barley (*Hordeum vulgare*), triticale, ryegrass (*Lolium* spp.), sorghum (*Sorghum bicolor*), buckwheat (*Fagopyrum esculentum*), pearl millet (*Pennisetum glaucum*), rye (*Secale cereale*) and oat (*Avena sativa*). Some chapters have also included a discussion of the potential of protoplast manipulations and genetic transformation of the particular crop under discussion.

Our thanks to authors for their time and effort to write these chapters and to the reviewers for working to improve the quality of the manuscripts.

S. Mohan Jain
S.K. Sopory
R.E. Veilleux

# Acknowledgements

Ever since I finished my Ph.D. on *In vitro* haploids in higher plants, I have wanted to edit a book on this important subject and tucked this thought in the back of my mind. On separate opportunities, I mentioned it to Profs. S.K. Sopory and Richard E. Veilleux, inviting them to co-edit such a book. They graciously accepted my invitation to become co-editors. Their critical review of manuscripts and valuable suggestions have substantially improved the quality of these volumes. I am thankful to both Sudhir and Richard for helping me on this ambitious project and it has been a great pleasure working with them.

I appreciate the invited authors for their punctuality in meeting deadlines for submission of their contributions and all the reviewers (named in the General Preface) for constructive and timely critical reviews of the manuscripts. Their comments have been extremely useful for improving the quality of these volumes.

I am thankful to my colleagues Prof. Eija Pehu, Mr. Tapio Poutala, and Mr. Matti Teittinen for their assistance.

While editing this book, I had the opportunity to visit the University of Tuscia, Italy, as a visiting professor fellow. I am thankful to Prof. Eddo Rugini for his warm hospitality. During my short stay, I managed to find the time to edit several manuscripts.

Also, with great love and affection, I want to thank my daughters Sarita and Sonia, and my wife, Marja Liisa, for their unceasing patience and understanding while I was working on this time-consuming project.

Finally, I express my deepest sense of appreciation to Adrian Plaizier of Kluwer Academic Publishers, the Netherlands, for giving us the opportunity to work on this project. Adrian has always been cooperative and helpful, encouraging me with intelligent advice.

Book Project Leader
S. Mohan Jain

# 1. Haploidy in rice

S.S. GOSAL, A.S. SINDHU, J.S. SANDHU, RAMAN SANDHU-GILL, BALDEV SINGH, G.S. KHEHRA, G.S. SIDHU and H.S. DHALIWAL

**Contents**



## 1. Introduction

Rice (*Oryza sativa* L; $2n = 2x = 24$) belongs to family Gramineae and the genus *Oryza* which includes 24 recognized species, of which 22 are primitive and the remaining two, *O. sativa* and *O. glaberrima* Steud., are cultivated. All the rice cultivars grown in Asia, Europe and America are *O. sativa*, whereas many of those cultivated in West Africa are *O. glaberrima*. *O. sativa* consists of two distinct land races, Indica and Japonica. Indicas are grown in the humid tropics whereas Japonicas are cultivated in Japan and Northern China under cooler climate. The distinction between Indica and Japonica is partly based on the existence of reproductive barriers between them (Oka, 1988). These races differ with respect to genes and characters that are associated non-randomly; no single gene representative of Indica-Japonica differentiation can be identified (Sato *et al.*, 1986; Glaszmann, 1987). On the basis of isozyme (Glaszmann, 1987) or RFLP patterns (Wang & Tanksley, 1989), these groups can now be further classified into six distinct varietal groups (VG). The Japonica and Javanica races are in VG6. The intermediate varieties previously referred as Indicas can be differentiated from true Indicas of VG1 and belong to the other varietal groups.

Today, rice is the principal food crop in the world, with more than 90% of the world's rice produced and consumed in Asia. More than half of Asian

*S.M. Jain, S.K. Sopory & R.E. Veilleux (eds.),* In Vitro *Haploid Production in Higher Plants, Vol. 4,* 1–35.

rice is produced in India and China. World rice production has increased faster than human population growth during the last three decades. However, a yield plateau seems to have already been reached but further increases in rice production are needed to meet the demand of the ever-rising population pressure. World annual rice production must be increased from 527 m tonnes today to 556 m tonnes by 2000 and 758 m tonnes by 2020 (IRRI, 1989). Rice is naturally self-pollinating and rice cultivars tend to be pure lines, although there has been considerable interest in the development of hybrid rice due to a 10–15% yield increase over conventional cultivars. An efficient mechanism to ensure cross-pollination in hybrid seed production has been a major obstacle (McKenzie *et al.*, 1987).

Increase in rice production has mainly been due to the development of high yielding, semi-dwarf, photoperiod-insensitive, pest resistant cultivars. To meet the global demand for rice consumption, there is an urgent need for rapid development of even higher yielding rice. Biotechnology offers great potential for creation, conservation, characterization and utilization of germplasm for rice breeding programmes. By using biotechnological techniques, such as anther culture, embryo rescue, and somaclonal variation, several new rice cultivars have already been released. Use of anther culture has contributed more than 100 new rice cultivars in China (Meifang, 1992), and over 42 Japonica rice cultivars in Korea. Exploitation of gametoclonal variation has resulted in the development and release of cv. Dama in Hungary (Heszky *et al.*, 1991).

Production of doubled haploids through anther culture is a rapid approach to homozygosity that shortens the time required for the development of new rice cultivars compared to conventional methods which require at least 6–7 generations. Haploids are also valuable to detect and fix desirable recessive traits introduced through mutation or hybridization. Moreover, anther and microspore culture of $F_1$ hybrids lead to fixation of gene combinations which otherwise may not be possible to isolate from a segregating population through conventional means. An ideal situation would be the rapid fixation of heterotic gene combinations from an $F_1$ generation to develop homozygous lines as good as heterotic $F_1$ hybrids (Fig. 1).

Hu and Zeng (1983) compared three breeding methods in rice, viz. anther culture, bulk method, and pedigree method. Among the three methods, anther culture was ideal for developing homozygous lines in the shortest time, i.e., in only four generations. The doubled haploids (DH) produced from the $F_1$ by anther culture increase selection efficiency, especially when dominance variation is significant. In conventional breeding, early generation lines exhibit variation due to both additive and dominance effects whereas DH lines derived by anther culture of an $F_1$ will show only additive variance; therefore, high heritability can be expected due to the elimination of dominance effects. Thus, compared to an $F_2$ population, fewer DH plants are needed to select desirable recombinants. Baenziger & Schaeffer (1983) calculated that in a cross segregating for three recessive genes, 1/8 of the

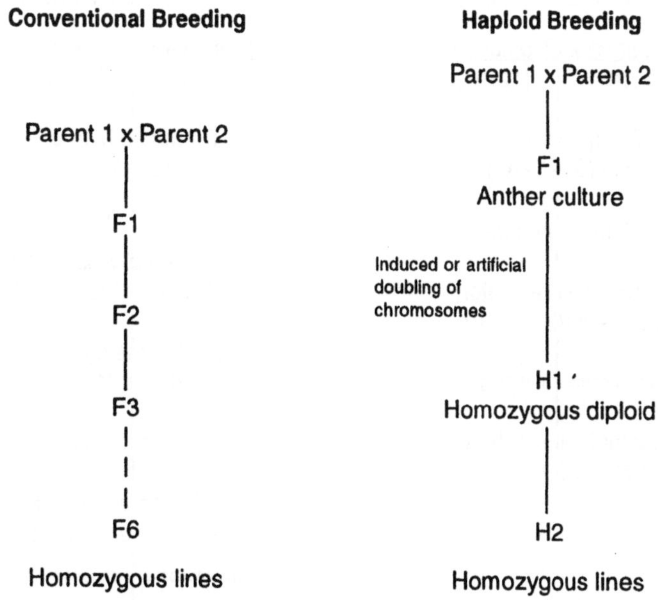


*Figure 1.* Development of homozygous true breeding lines through conventional breeding and anther culture (haploid) breeding.

DH derived from the $F_1$ would be selected compared to 1/64 of the $F_2$. Similarly, if selecting for three dominant genes, 1/8 of the DH would be selected and would breed true compared to 27/64 of $F_2$ plants selected on the basis of phenotype, of which only 1/64 would breed true. Shen *et al.* (1983) estimated that about 150 pollen plants derived from $F_1$ anthers instead of 4000–5000 $F_2$ plants would be sufficient for selecting desirable genotypes. By selection of DHs produced from either an $F_1$ hybrid or a promising segregant, desirable recombinants can be selected and no segregation in successive generations is expected.

Anther culture has potential application in wide crosses also. Chu (1982) outlined a scheme involving anther culture for the development of substitution and addition lines. This could be of immense use in the transfer of desirable genes from alien sources across the barriers of genomic incompatibility (Fig. 2).

Anther culture can save time, labour and money in achieving specific goals in rice breeding. A cost/benefit analysis made by Sanint *et al.* (1992) has indicated that anther culture could reduce the cost by 60,000 to 200,000 US$ per cultivar depending upon the genotype/ecosystem targeted. The DH approach is increasingly being used for rapid development of mapping populations and construction of genetic linkage maps for traits of interest.

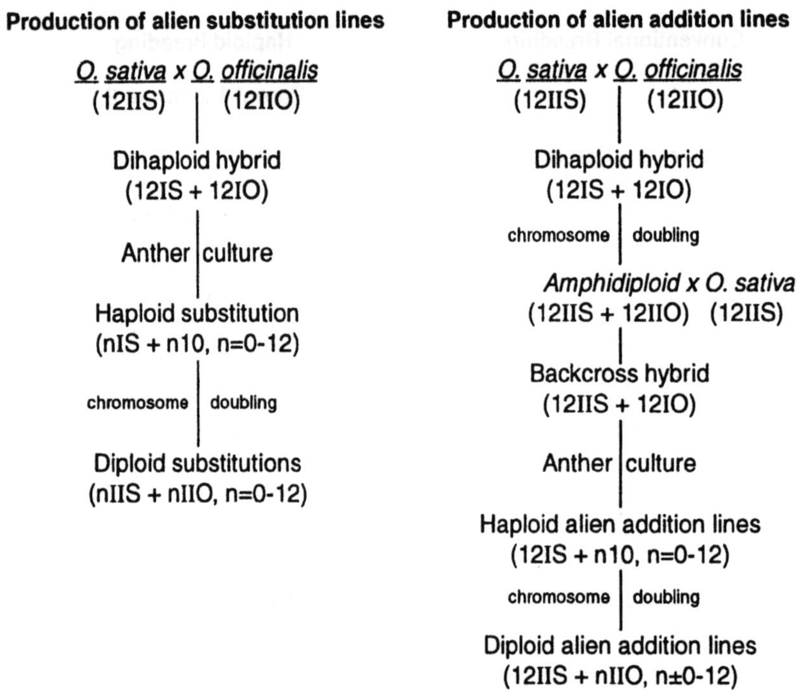


*Figure 2.* Production of alien substitution and addition lines through anther culture.

## 2. Anther culture

*In vitro* haploid production by anther culture was first developed at the Delhi University by Guha & Maheshwari (1964, 1966) in *Datura*. Later, this technique was used in rice by Japanese (Niizeki & Oono, 1968; Nishi & Mitsuoka, 1969), Korean (Harn, 1969), and Indian researchers (Guha *et al.*, 1970). Production of haploid/doubled haploid rice plants has since been extended to a diverse array of rice genotypes (Table 1). Recently, emphasis has been given to Indica type cultivars and efficient protocols for anther culture (Reddy *et al.*, 1985; Shahjahan *et al.*, 1985; Zapata & Torrizo, 1986; Zapata *et al.*, 1986a; Mercy & Zapata, 1987; Kavi Kishor *et al.*, 1989; Datta *et al.*, 1990a; Ogawa *et al.*, 1992; Lenka & Reddy, 1994).

### 2.1. *Procedure*

Boots are collected from the field when the distance between the base of the flag leaf and the auricle of the last leaf is 3–8 cm, depending upon the cultivar. The middle part of the collected panicle contains microspores at the uninucleate stage. The collected boots are wrapped in plastic bags and kept in the refrigerator for cold treatment at 4°C for 7–10 days. Panicles are carefully isolated by opening the flag leaf with the help of a sterilized needle.

*Table 1.* Anther, microspore and ovary culture for production of haploids/doubled haploids in rice

| Plant material | Explant/stage | Medium | Remarks | Reference |
|---|---|---|---|---|
| *O. sativa* | Anthers | Blaydes 1966 medium + NAA (9.4μM) + kin (10 μM) | Obtained 0.57% regeneration after 4-8 wks in culture | Niizeki and Oono (1968) |
| 20 Indica cultivars | Anthers with uninucleate pollen | Medium supplemented with IAA, 2,4-D, YE and CW | Observed variation between cultivars for androgenic response. Highest callus was observed in 'Assam 5' | Guha *et al.* (1970) |
| Keng, Keng × Shien $F_1$ hybrids | Anthers | $N_6$ or SK-8 liquid medium | Callus induction from unreleased pollen in both media | Chen and Chen (1979) |
| *O. sativa* | Anthers with binucleate to trinucleate microspores<br><br>Mature ovary | Miller's macro+micro+Morel's vit. + NAA (3mg $l^{-1}$) + YE (2000 mg $l^{-1}$) + sucrose (6% w/v) | Obtained green haploid and mixoploid plants | de Beauville (1980) |
| *O. sativa* | Anthers with late uninucleate to binucleate microspores<br><br>Uninucleate to mature ovary | 1. $N_6$+ MCPA (0.125 mg $l^{-1}$) + sucrose (3% w/v)<br>2. $N_6$ or MS + IAA (0.5 mg $l^{-1}$) + kin (1-2 mg $l^{-1}$) + NAA (0.25 mg $l^{-1}$) + sucrose (3% w/v) or $N_6$+ MCPA(0.033 mg $l^{-1}$) + sucrose (3% w/v) | Green as well as albino haploids were obtained; float culture of unhusked flowers was used | Zhou and Yang (1981[a,b,c]) |

*Table 1.* Continued

| Plant material | Explant/stage | Medium | Remarks | Reference |
|---|---|---|---|---|
| *O. sativa* | Anthers with late uninucleate to early binucleate microspores<br>Uninucleate to mature ovary | 1. $N_6$+ 2,4-D (2 mg $l^{-1}$) + CH (500mg $l^{-1}$) + sucrose (4% w/v)<br>2. $N_6$ + CH (500 mg $l^{-1}$) + kin (2 mg $l^{-1}$) + NAA (0.05 mg $l^{-1}$) + sucrose (3% w/v) | Green haploid, diploid and some albino plants | Kuo (1981) |
| Minehikari (Japonica) | Anthers with mid to late uninucleate microspores | R-3 (modification of LS) + sucrose (4% w/v) + various combinations of NAA, IAA and 2,4-D + agarose (0.2% w/v or 0.4% w/v) | Sucrose concentration of 4% w/v was found to be beneficial for callus induction in both 0.4% w/v and 0.2% w/v agarose; addition of 50$\mu$M mannitol increased the callus induction frequency | Chaleff and Stolarz (1981) |
| Calrose-76, Labelle, Mars, Nortai, Starbonnet, Ma | Anthers with mid to late uninucleate microspores | 1. Blaydes salts + YE (1g $l^{-1}$) + CM (150 ml $l^{-1}$) + IAA (2mg $l^{-1}$) + sucrose (3% w/v)<br>2. Blaydes salts + CM (500mg $l^{-1}$) + YE (250mg $l^{-1}$) + 2,4-D (1 mg $l^{-1}$) + NAA (0.5 mg $l^{-1}$) + kin (1 mg $l^{-1}$) + sucrose (3% w/v)<br>3. MS + glutamine ($10^{-3}$M) + inositol (100mg $l^{-1}$) + IAA (1mg $l^{-1}$) + kin (1mg $l^{-1}$) + sucrose (3% w/v) | Obtained plants in 'Calrose 76' and observed variability in the regenerated plants for height, seeds, flowering time, awn length, seed weight, etc. | Schaeffer (1982) |
| Giza 170, Taipei 309 | Anthers with mid | 1. $E_{10}$ + sucrose (2% w/v) + glucose (0.5% w/v) + BA | Used conditioned medium 1:8 dilution and obtained 1.6 | Zapata *et al.*, |

| | | | | |
|---|---|---|---|---|
| | uninucleate to early binucleate microspores | (0.5 mg $l^{-1}$) + 2,4-D (1 mg $l^{-1}$)<br>2. MS + NAA (1 mg $l^{-1}$) + kin (1mg $l^{-1}$) | times higher callus than control for 'Taipei 309' while conditioned medium reduced the callus induction as well as green plant regeneration in 'Giza 170' | (1985) |
| Indica cultivars ARC 1557, Tetep, Mohini, Pankaj, Govindobhoa, Radhunipagal Small fruits, Meghna | Anthers with uninucleate microspores | 1. He-2 + 2,4-D (9 μM) or He-5 + sucrose (5% w/v) or $N_6$ or $R_3$<br>2. He-2 + different concentrations and combinations of 2,4-D, NAA, picloram, zeatin and kin + glucose (5% w/v)<br>3. He-2 + 2,4-D (9 μM) + sucrose (3,5,7,9 or 11%)<br>4. MS + NAA (1.5 μM) + kin (9.5μM) | Different cultivars responded differently to different media. He-2 and He-5 media induced callus at a higher frequency than R-3 and $N_6$. Medium with 2,4-D (9 μM) picloram (0.3μM) and zeatin (0.5 μM) was most effective for callus induction and green plant regeneration. Best results were obtained when anthers treated at 35°C for 5 min followed by 10°C for 7 days. An increase in sucrose concentration from 3 to 5% in the culture medium resulted in an increase in percentage of callus induced. | Reddy *et al.*, (1985) |
| 3 Indica, Indica x Japonica 4 $F_2$ lines | Anthers | $N_6$, modified $N_6$, $H_5$, modified $H_5$ $SK_8$ | The response of different cultivars was different in 5 media indicating existence of genotypic variability. Green plant regeneration was higher in calli induced on $H_5$ medium. | Shahjahan *et al.*, (1985) |
| Indica cultivars IR-42 | Anthers with mid uni-nucleate to early binucleate microspores | $B_5$+ 2,4-D (1 mg $l^{-1}$) + BA (0.5mg $l^{-1}$) + IAA (0.5 mg $l^{-1}$) or $B_5$+ zeatin (1 mg $l^{-1}$) + 2,4-D (2 mg $l^{-1}$) + picloram (0.07 mg $l^{-1}$) | Heat treatment at 35°C for 15 min increased significantly the callus induction in both media | Zapata and Torrizo (1986) |
| Basmati 370 | Anthers with uninucleate to | 1. $B_5$+ 2,4-D (1 mg $l^{-1}$) + NAA (0.5mg $l^{-1}$) + BA | 20 K rad treatment of seeds of donor plants reduced the | Zapata *et al.*, |

*Table 1.* Continued

| Plant material | Explant/stage | Medium | Remarks | Reference |
|---|---|---|---|---|
| | binucleate microspores | (0.5mg $l^{-1}$) + sucrose (2% w/v) with or without glucose (0.5% w/v) (liquid medium)<br>2. MS + kin (1mg $l^{-1}$) + NAA (1 mg $l^{-1}$) + ABA (20 mg $l^{-1}$) for 4 weeks (semi-solid medium)<br>3. ABA free medium. | callus induction but also reduced the percentage of albino plant regeneation<br>No green plants were regenerated in control; addition of glucose 5 g $l^{-1}$ improved the callus induction frequency | (1986) |
| Taipei-309 (Japonica), IR-5, Basmati 370, Pankaj | Anthers | 1. $E_{10}$+ 2,4-D (1 mg $l^{-1}$) + BA (0.5 mg $l^{-1}$) + IAA (0.5 mg $l^{-1}$) + sucrose (2% w/v) + glucose (5g $l^{-1}$)<br>2. MS + NAA (1 mg $l^{-1}$) + kin (1 mg $l^{-1}$) + sucrose (3% w/v) | 60% of the callus forming anthers were plated on edge and 40% were flat. Most of the anthers plated flat callused on both lobes but callusing on one lobe was also observed. Most of the anthers plated on edge produced callus on the upper lobe. | Mercy and Zapata (1987) |
| Taipei-309 (Japonica) | Uninucleate microspores | $E_{10}$<br>$E_{10}$ + alanine (1mM)<br>$E_{10}$ + glutamine (1mM)<br>$E_{10}$ + proline (1mM) | Glutamine and proline had stimulatory effects on pollen culture. Only large 50-58$\mu$M diam with thin pink coloured cell walls produced pollen embryos. Division of 40$\mu$M diam thick-walled pollen was not observed. | Cho and Zapata (1988) |
| PR106 × Jaya)<br>*O. sativa* x *O. rufipogon*<br>*O. sativa* x *O. ongistaminata* | Anthers | Potato-2 or $N_6$+ NAA or 2, 4-D (2 mg $l^{-1}$) + kin (1 mg $l^{-1}$) + sucrose (6% w/v) + agar (0.75% w/v) | 1. Potato-2 medium was superior to $N_6$<br>2. Early callusing when NAA was present instead of 2,4-D<br>3. *O. sativa* x *O. longistaminata* gave no response except their $F_2$ progeny formed callus from anthers. | Rout *et al.*, (1989) |

| | | | | |
|---|---|---|---|---|
| 17 Japonica, 31 Indica, 2 Indica x Japonica and one non classified cv. | Anthers | 1. $N_6$ + NAA (1 mg $l^{-1}$) + agar (0.8% w/v)<br>2. MS + kin (3 mg $l^{-1}$) + NAA (0.5 mg $l^{-1}$) | Japonica cultivars were more responsive than Indicas; 27% of the plants obtained were albinos. | Reiffers and Freire (1990) |
| $F_1$ heterotic hybrids | Anthers | MSN or SK-1 with 2,4-D (0.7 mg $l^{-1}$) + NAA (2.5 mg $l^{-1}$) + kin (0.75 mg $l^{-1}$) + sucrose (4.% w/v) | MSN medium was better in all the $F_1$ hybrids while regeneration frequencies were higher in calli raised on SK-1 medium. | Raina *et al.*, (1989) |
| Indica cultivars: PTB-33, Rasi, Crossa, Pokkali, Tellahemsa, Yerregaluvadlu, Mahsuri<br>Indica x Japonica: TCA-2 x YVSD, Basmati $D_5$ x Basmati, Mahsuri x Bala, TN-1 x Carreon | Anthers with early to mid uninucleate microspores | 1. $N_6$, He-2, He-5, LS, or $R_2$ with different concentration and combinations of 2,4-D, NAA, kin and sucrose (2,3,4 or 5% w/v)<br>2. $R_2$ + NAA (2 mg $l^{-1}$) + kin (2 mg $l^{-1}$)<br>3. LS + NAA (1 mg $l^{-1}$) + kin (4 mg $l^{-1}$) | 1. Variation for callus induction and green plant regeneration was found.<br>2. $R_2$ with NAA (2 mg $l^{-1}$) + kin (0.5 mg $l^{-1}$) was found suitable for callus induction.<br>3. 3% w/v sucrose gave better results. | Kavi Kishor *et al.*, (1989) |
| WU 1OB | Anthers | 1. $N_6$+ 2,4-D (2 mg $l^{-1}$) + sucrose (6% w/v) + gourd stem exudate (0, 10, 15, 20 or 25% w/v).<br>2. MS + kin (2 mg $l^{-1}$) + NAA (0.5 mg $l^{-1}$) + BA (0.5 mg $l^{-1}$) | Callus response was highest from main tiller. However, regeneration frequencies on calli basis did not differ except for tertiary tiller. | Hadi and Raina (1989) |
| Indica and Japonica | Anthers | 1. $N_6$ + 2,4-D (2 mg $l^{-1}$) + sucrose (6% w/v) + gourd | Addition of gourd stem exudate (15-20% w/v) lead to | Chengmei and |

*Table 1.* Continued

| Plant material | Explant/stage | Medium | Remarks | Reference |
|---|---|---|---|---|
| cultivars | | stem exudate (0, 10, 15, 20 or 25% w/v).<br>2. MS + kin (2 mg $l^{-1}$) + NAA (0.5 mg $l^{-1}$) + IAA (0.5mg $l^{-1}$) | early callus induction and also increased the callus induction frequency in both Japonica and Indica rices and also had increased plant regeneration. | Zhenghua (1990) |
| Taipei-309 (Japonica) IR-43] Indica IR-36] | Anthers with late uninucleate microspores | $E_{10}$+ 2,4-D (1 mg $l^{-1}$) + BA (0.5 mg $l^{-1}$) + IAA (0.5 mg $l^{-1}$) + sucrose (20 g $l^{-1}$) + glucose (5 g $l^{-1}$) + alaline (1mM) or glutamine (1mM) or proline (1mM) | Addition of proline and glutamine reduced the albinism frequency; 72% of regenerated plants were homozygous diploids | Zapata *et al.*, (1990) |
| Japonica x Japonica, Japonica x Indica crosses | Anthers | 1. $N_6$ + NAA (1 mg $l^{-1}$) + sucrose (6% w/v)<br>2. MS + NAA (0.5 mg $l^{-1}$) + kin (3 mg $l^{-1}$) | Variability for callus induction and plant regeneration was observed | Reiffers and Freire (1990) |
| Nagdongbyea (Japonica) Somgaughbyeo (Tongil types) | Anthers | One step method<br>$N_6$ + NAA (1 mg $l^{-1}$) + kin (4 mg $l^{-1}$) + agar (1.2% w/v)<br>Two step method<br>1. $N_6$Y-1 + NAA (2 mg $l^{-1}$) + kin (1 mg $l^{-1}$) + agar (0.8% w/v)<br>2. $N_6$Y-1 + IAA (0.2 mg $l^{-1}$) + kin (1 mg $l^{-1}$) + agar (0.8% w/v) | Single step anther culture can be effectively employed which can save time, energy and money in obtaining haploids directly from anthers | Chung *et al.*, (1991) |
| Taipei-309 | Anthers with mid uninucleate to early | 1. $B_5$ + 2,4-D (1 mg $l^{-1}$) + BA (0.5 mg $l^{-1}$) + IAA (0.5mg $l^{-1}$) + sucrose (2% w/v) | This procedure saved time (10-15 days); direct plantlets were produced on medium M-10 and P-2; % albino | Karim and Zapata (1990) |

| | | | | |
|---|---|---|---|---|
| | binucleate microspores | 2. $M_{10}$ + NAA (1.5mg $l^{-1}$) + BA (0.5 mg $l^{-1}$) + IBA (0.5 mg $l^{-1}$) + kin (0.75mg $l^{-1}$) + sucrose (6% w/v)<br>3. $P_2$ + NAA (2 mg $l^{-1}$) + kin (0.5 mg $l^{-1}$) + sucrose (6% w/v) | plantlets was greater on medium P-2 | |
| Indica and Japonica | Microspores | MS + glutamine + CH + Ficoll (10% w/v) | Green plants obtained in both Indica and Japonica cultivars | Datta *et al.*, (1990a) |
| Indica cv. IR-24 | Mechanically isolated pollen grains (42-50 μM) | $R_2$ (1/4 strength) + $B_5$ vitamins + 2,4-D (0.5 mg $l^{-1}$) + sucrose (5mM) + mannitol (0.4M) | Green and albino plants were regenerated; a cold treatment of 25 days gave the highest colony formation | Ogawa *et al.*, (1992) |
| Three $F_1$ hybrids of Japonica cultivars | Anthers | 1. $N_6$ medium with hormones in disposable petridishes or in glass jam jars<br>2. MS + kin (2 mg $l^{-1}$) + NAA (0.5 mg $l^{-1}$) + sucrose (4% w/v) | Regeneration of green plants was better in jam jars in all genotypes | Courtois (1992) |
| Japonica x Japonica<br>Japonica x Indica | Anthers with uninucleate microspores | 1. $N_6M$ or $N_6AK$ supplemented with MCPA, kin, NAA with sucrose (4.5% or 5.5% w/v)<br>2. For Indica $R_2$ + kin (2 mg $l^{-1}$) + NAA (0.5 mg $l^{-1}$) + sucrose (6% w/v) or for Japonica $R_6$ + kin (1 mg $l^{-1}$) + NAA (0.1 mg $l^{-1}$) + sucrose (4% w/v) | With an increase in time in culture, the ratio of green/albino plantlets did not change; however, the overall plant regeneration frequency was affected with the time in culture | Guiderdoni *et al.*, (1992) |
| Indica rice cultivar, $F_1$ and | Anthers with uninucleate | $N_6$ + 2,4-D (2 mg $l^{-1}$) + sucrose (3% w/v) | Addition of tryptophan (25 mg $l^{-1}$), use of mannitol (1% | Gosal *et al.*, |

*Table 1.* Continued

| Plant material | Explant/stage | Medium | Remarks | Reference |
|---|---|---|---|---|
| $F_2$ resulting from high yielding Indica rice x Basmati cultivars | microspores | | w/v) along with sucrose (3% w/v), cold pretreatment (8°C for 7 days) and culturing anthers from primary tillers enhanced anther callusing. | (1991, 1993); Sandhu *et al.*, (1993[a,b]) |
| Indica cultivars CR666-34-3, SR 26-B, Basmati Sd-10, Ptb-33 | Anthers | 1. $R_2$, $N_6$ or MS with different combinations of 2,4-D and NAA<br>2. $R_2$, $N_6$ or MS with different combinations of IAA, NAA, BA and kin with sucrose (3-6% w/v) | CR 666-34-3 and SR-26-B were poorer responding than Basmati Sd-10 and Ptb-33.<br>Early callus induction was obsrved in media having NAA;<br>$R_2$ and $N_6$ were better callus induction media; MS media supplemented with 6% w/v sucrose gave higher green plant regeneration | Lenka and Reddy (1994) |
| Mediterranean Japonica rice cv. Miara | Anthers with uninucleate microspores | $N_6$ AK + NAA (2 mg $l^{-1}$) + kin (0.5 mg $l^{-1}$) + agarose (0.55% w/v) + sucrose (6% w/v) | Plating of rice anthers on colchicine (250-500 mg $l^{-1}$) containing medium for 24-48 h resulted in 1.5-2.5-fold increase in doubled haploid green plants | Alemanno and Guiderdoni (1994) |

Each spikelet is cut from the base so that the filament of the anther is also cut and then spikelets are tapped gently at the mouth of the test tube/flask/petri dish in order to release anthers directly onto the medium. Anthers form callus after 40–45 days, and calli (1–2 mm) are placed on regeneration medium for complete plant regeneration.

There are two pathways in rice androgenesis which lead to pollen embryogenesis (Sun, 1981; Yang & Zhou, 1979):

*Pathway A*: The first pollen division is unequal and forms two nuclei – vegetative and generative – which differ in size and stainability (Chen, 1977; Sun, 1981; Yang & Zhou, 1979). The vegetative nucleus further divides and forms callus/embryos while the generative nucleus degenerates. In some cases, the generative nucleus may also divide a few times (Sun, 1981).

*Pathway B*: The first pollen division is equal leading to two similar nuclei (Cheng, 1978).

## 3. Factors affecting anther culture

There are several factors, such as genotype, physiological state of the donor plant, physiological stage of the microspores, culture medium, growth regulators, sucrose, and shock pretreatments, that affect the response of anther/-microspore culture for producing androgenic callus and plant regeneration (Table 1). Mercy & Zapata (1987) studied the effect of anther orientation and found that most of the anthers plated flat against the medium callused on both lobes; however, anthers plated on edge produced callus on the upper lobe. Toriyama & Hinata (1985) obtained anther calli after one month of culture by culturing panicle segments with florets at the uninucleate microspore stage whereas, by cutting the tip of each floret, callus induction occurred within three weeks. Sohn *et al.* (1987) also succeeded in obtaining callus from panicle culture.

To save time, labour and money, several researchers have developed a single step culture method for plant regeneration on embryo/callus induction medium without transfer to regeneration medium. Karim & Zapata (1990) obtained plantlets directly from cultured anthers of "Taipei 309". Similarly, Chung *et al.* (1991) developed a system for direct plantlet regeneration from cultured anthers of a Japonica cultivar. Marassi *et al.* (1993) regenerated green plants using a one-step culture procedure on eight different rice cultivars.

## 4. Genotype

In general, Japonica rice cultivars respond better than Indica cultivars (Chen & Lin, 1976; Hu, 1981; Rush & Shao, 1982; Chu, 1982; Woo & Chen, 1982; Zapata *et al.*, 1986b; Quimio & Zapata, 1990) and sometimes $F_1$ hybrids perform better than their inbred parents (Chen, 1986a,b). Genotypic dependence of androgenic response of different rice germplasm is summarized in Table 1.

Iyer & Raina (1972) demonstrated that, of 15 cultivars studied, only five showed androgenic response and only one regenerated roots and shoots. Guha-Mukherjee (1973) found that some Indica and Japonica cultivars did not respond to anther culture, but cv. Assam 5 formed callus in 26% of cultured anthers. Similarly, Oono (1975) reported variation in callus formation capability among the lines tested from 0.5 to 8.0%, and regeneration of albino and green plants was also genotype-dependent. Similar results were obtained by Chaleff (1980) indicating that Indica cultivars regenerated fewer green plants (only 5%) than Japonica types (up to 90%). Hu (1985) regenerated green plants from 10% of cultured anthers of Japonica cultivars compared to 1% for Indica rice. Among 11 Indica cultivars examined by Aruna & Reddy (1988), the frequency of green plants regenerated varied from 0 to 47.5%.

Similarly, the callus forming ability from rice anther culture and time required for callus induction have been genotype-dependent (Abe, 1992; Reddy *et al.*, 1985). Kim *et al.* (1991) reported the best response from Japonica × Japonica hybrids followed by Indica × Japonica and then Indica × Indica crosses. Several other researchers have also noticed a decline in androgenesis in this order: Japonica > Japonica × Indica > Indica (Chen & Lin, 1976; Tsai & Lin, 1977; Chaleff, 1978; Miah *et al.*, 1985). In an examination of the androgenic response of three Japonica cultivars and their isogenic male sterile lines, the male sterile lines were either less responsive or not different from the original cultivars depending on the time of microspore degeneration of the male steriles (Courtois & Taillebois, 1990). Among Japonica × Indica hybrids, both callus induction and green plant regeneration have varied considerably depending upon the specific cultivars used to construct the hybrids (Narasimman & Rangasamy, 1993).

We have studied various factors affecting anther culture success in Indica rice. Major attention has been focused on commercial high-yielding Indica rice and Basmati cultivars and their hybrids (Gosal *et al.*, 1991, 1993; Sandhu *et al.*, 1993a,b). In crosses involving Basmati as one parent, a higher frequency of green plants was regenerated. The doubled haploid lines produced are under field evaluation (Fig. 3).

*Figure 3.* Field evaluation of anther-derived plants/lines resulting from crosses involving high yielding Indica and Bamati cultivars. (1) Anther-derived plants grown in the field. (2) Evaluation of doubled haploid lines in replicated trials.

## 5. Genetics of anther culturability

Callus induction ability is inherited as a recessive character conditioned by a single block of genes and Japonicas appear to be good combiners for callus induction (Miah *et al.*, 1985). Quimio & Zapata (1990) concluded, from a diallel analysis of anther culturability and green plant regeneration, that the effects of genotype and genotype × medium interactions were significant. Combining ability analysis suggested a predominance of additive gene effects in controlling both characters, with Japonica cultivars having higher combining ability for green plant regeneration. Reciprocal effects were not significant, ruling out maternal control of these characters. Regeneration was partially dominant with high response recessive and controlled by only a few genes. Interallelic interactions were not found. Chen & Chen (1993) dissected the androgenic response into four components: number of anthers producing at least one callus, frequency of callus induction, frequency of callus differentiation, and culture efficiency (number of differentiated calli divided by the total number of anthers cultured). They found that a model using both additive and dominance genetic variance could explain the inheritance of the four traits with additive greater than dominance effects. Thus, androgenic response is controlled by genes which can be introduced by crossing into otherwise unresponsive genotypes. Anther culture of anther-derived plants from an $F_1$ between *O. glaberrima* and *O. sativa* did not result in increased callus induction or plant regeneration (Woo *et al.*, 1983).

## 6. Physiological state of donor plants

The physiological state of the donor plants has a considerable impact on the response of *in vitro* anther/microspore cultures because initially plants tend to conserve resources under non-optimal growth conditions, often resulting in pollen abortion. Therefore, a primary requirement is to control the growth conditions of donor plants to ensure the quality of microspores. Early flowering plants with numerous flowers generally produce viable microspores that are more responsive to *in vitro* culture. Many factors that affect plant growth, viz. plant age, photoperiod, light intensity, temperature, season, growing conditions, and nutritional status, also affect the anther culture response. Hu *et al.* (1981) obtained best anther culture response when plants were grown under bright sun at 18–20°C. Raina *et al.* (1987) suggested that Indica cultivars have a specific need of high temperature and gave best anther culture response when the plants were grown at 23.2–34.2°C compared to 16.4–29.1°C.

Anther culture response of Indica rice cultivars was affected by the photoperiod and thermoperiod under which the plants were grown (Sun *et al.*, 1992). Callus induction rate was highest when donor plants were grown under mid or short photoperiod and at mid or low temperatures. The highest rate of callusing was 42.2% obtained at 25.7°C under 14 h photoperiod. The ratio of green to albino plantlets was also higher under a short photoperiod. Ling *et al.* (1990) found no difference in callus-forming ability of photoperiod-sensitive and insensitive rice genotypes. However, Huang *et al.* (1983) reported that exposure of donor plants to high temperature had an adverse effect on green plant regeneration.

The application of growth substances and certain feminizing agents has improved anther culture response. A 50 ml solution of 500 ppm Alar, a feminizing agent, applied daily for seven consecutive days at the base of 3–5 cm inflorescences of donor plants, increased the response up to 2.4 fold (Zapata *et al.*, 1990). Wang *et al.* (1979) reported enhancement in anther culture response with 4000 ppm ethrel (2–chloroethyl phosphonic acid) treatment for 48 h at 10°C whereas Chaleff & Stolarz (1982) could not improve anther response in ethrel-treated donor plants. The application of male gametocides before or after meiosis has also enhanced callusing frequency and earliness in callusing (Beaumont & Courtois, 1990a,b; Zhao & Qi, 1991). Chen & Lin (1976) observed that anthers collected and cultured at the beginning of the flowering period were more productive than those cultured at the end of the flowering period. Furthermore, Chen & Tsay (1984) reported that the frequency of callus formation was significantly lower in anthers collected from the tertiary tiller than in those from primary and secondary tillers. Radiation treatments to seed of plants destined for anther culture have also been attempted. Aldemita & Zapata (1991) treated seeds of several rice cultlivars with 1.82 kr $min^{-1}$ and found that radiation treatments enhanced callus induction and plant regeneration of recalcitrant cultivars but decreased both responses of normally responsive cultivars.

## 7. Developmental stage of the microspores

The microspore developmental stage is the most critical factor that affects the frequency of pollen embryo/callus formation. Success depends upon accuracy in selecting panicles/buds containing the appropriate stage of the microspores. There are a number of stages at which microspores can be diverted into embryogenesis and these stages vary with the species (Sunderland & Dunwell, 1977). In rice, the optimal stage of microspore development for anther culture response occurs only briefly. Tsay & Chen (1984) observed the induction process which diverted microspore development to the sporophytic pathway lasted 5–6 days in cultured anthers and could be extended by cold treatment. Chen (1977) found that maximal response occurred at the mid-uninucleate microspore stage; before or after this stage, response declined sharply. The ability of callus to regenerate green plants was also determined by the microspore stage. Calli derived from microspores at more advanced stages exhibited a lower capacity for plant regeneration and the regenerated plants exhibited a higher frequency of albinos (Chen, 1977; Genovesi & Magill, 1979). Several researchers have successfully cultured anthers containing microspores at early uninucleate to early binucleate stages (Tian & Chen, 1983a,b; Zapata *et al.*, 1986a; Zapata & Torrizo, 1986; Schaeffer, 1982; Kavi Kishor *et al.*, 1989; Karim & Zapata, 1990). Zapata *et al.* (1986a) reported better success with anthers containing mid-uninucleate and early binucleate stages. A shift of maximal response from mid-uninucleate to early uninucleate stage was observed when sucrose concentration was increased from 6 to 9% (Chen, 1978).

The collection of boots from the field for anther culture requires some skill in order to select the right microspore developmental stage. Cytological examination is the most accurate way to identify the stage (Chen, 1976, 1977). Samples stained in iron alum haematoxylin have been used by Gupta & Borthakur (1987) for accurately identifying the developmental stage of microspores to improve the response to anther culture. But these procedures are too tedious and time consuming for mass anther culture in a breeding programme. Other criteria can be used, however, including: (1) the distance (3–6 cm) between the ligules of the flag leaf and next lower leaf (Oono, 1975; Chen & Chen, 1979; Chaleff & Stolarz, 1981); (2) colour and size of spikelets and anthers (Chen & Chen, 1979); and (3) texture of spikelets. These morphological criteria may vary among cultivars; Mercy & Zapata (1987) found that the most efficient auricle distance for callus induction was 7.0 cm for "Taipei 309" but 8.5 cm for "Kulu".

## 8. Culture medium

### 8.1. *Basal medium*

Culture medium is another important component for producing embryos/callus from cultured anthers/microspores. Nutrient requirements may differ for the induction of androgenesis and for growth of developing embryos (Wang *et al.*, 1974). Chu *et al.* (1975) developed $N_6$ medium with reduced $(NH_4)_2SO_4$ and increased $KNO_3$ specifically for anther culture of rice; this medium has subsequently been demonstrated to be the most suitable (Chu, 1981; Genovesi & Magill, 1979; Chen *et al.*, 1982; Tsay *et al.*, 1982). Both $B_5$ (Gamborg *et al.*, 1968) and modified LS (Linsmaier & Skoog, 1965) medium (R-3: Chaleff & Stolarz, 1981) which also have high concentrations of nitrate nitrogen and reduced concentrations of ammonium nitrogen have also been found to give good results (Chaleff & Stolarz, 1981; Zapata *et al.*, 1982).

$N_6$ medium has been superior to other media, viz. MS (Murashige & Skoog, 1962), potato extract, Blaydes (1966) SK-8 and Chaleff's R-2 (Chu, 1982; Lenka & Reddy, 1994). The ammonium nitrogen level in the medium was a crucial factor (Chu *et al.*, 1975; Chu, 1981; Gosal *et al.*, 1991, 1993; Sandhu *et al.*, 1993a) and its requirement by Indica cultivars is different from that of Japonica. The $N_6$ medium devised for Japonica cultivars was unsuitable for Indica varieties. For Japonica and Indica cultivars, the optimal $NH_4$ content was 7.0 me $l^{-1}$ and 3.0 me $l^{-1}$, respectively, but 4.8 me $l^{-1}$ for Japonica × Indica hybrids (Chen, 1983). Yang *et al.* (1980) used modified $N_6$ medium by replacing $(NH_4)_2SO_4$ and $KH_2PO_4$ with $NH_4HPO_4$ for Indica cultivars. Huang *et al.* (1978) modified $N_6$ medium and developed He-2 medium amended with half-strength ammonium, double-strength $KH_2PO_4$ and one-fiftieth the concentration of $MgSO_4$. Chen *et al.* (1986) observed a significant increase in anthers forming callus when 0.1 mM $Na_2Fe$ EDTA was added to the induction medium; however, further increase in the amount of iron (3.2 mM) was deleterious. Several modifications have been made to improve anther culture of Indica rice. He-2, He-5, potato-2, E-10, MSN modified media have been suitable for Indica rice anther culture (Raina *et al.*, 1989; Rout *et al.*, 1989; Reddy *et al.*, 1985; Zapata *et al.*, 1990). Shahjahan *et al.* (1985) found genotypic preferences for different media among Indica and Indica-Japonica hybrids. Zapata *et al.* (1990) also found genotypic differences for basal medium requirements, with $N_6$ medium giving better results in isolated pollen culture of "Taipei 309", a Japonica cultivar, whereas Indica cv. IR-43 responded better to $E_{10}$ medium, a modification of $B_5$ medium. Draz *et al.* (1991) studied the interaction of three different derivatives to $B_5$, $N_6$ and MS/$B_5$ basal media supplemented with different growth regulators and reported that there were significant differences among media as well as a significant interaction between cross and media. Callus induction medium also plays a role in inducing plant regeneration; for a particular callus induction medium, a specific regeneration medium may give a better response.

## 9. Plant growth substances

A combination of auxins [2,4–D (2,4-dichlorophenoxyacetic acid), NAA (napthaleneacetic acid), IAA (indole acetic acid)] cytokinins [BA (benzyladenine), kin (kinetin), zea (zeatin)], and organic substances [casein hydrolysate (CH), coconut water (CW), yeast extract(YE)] has been widely used in rice anther culture media (Niizeki & Oono, 1968; Harn, 1969; Guha *et al.*, 1970; Iyer & Raina, 1972). Several studies have shown that cytokinin and abscisic acid (ABA) were needed to obtain callus of high morphogenic potential (Chen, 1977; Chaleff & Stolarz, 1981; Inoue & Maeda, 1981; Lee & Chen, 1982; Torrizo & Zapata, 1986). Reddy *et al.* (1985) reported that NAA was better to induce callus, but 2,4–D was better for plantlet regeneration of Indica cultivars. However, high amounts of either NAA or 2,4–D enhanced callus formation but reduced plant regeneration. A modified medium with 2,4–D (9.0 $\mu$M), picloram (0.3 $\mu$M), and zea 0.15 $\mu$M) was most suitable for callus induction and green plant regeneration. The different effects of 2,4–D and NAA on callus growth may be genotype-dependent (Zhang, 1982). NAA (2 mg $l^{-1}$) induced a higher frequency of callus than 2,4–D in Japonica varieties and hybrids; however, plant regeneration was poorer for NAA-induced calli (Hu *et al.*, 1981). On the other hand, Rout *et al.* (1989) found that NAA improved both callus induction and plant regeneration. A combination of 1 mg $l^{-1}$ 2,4–D, 2 mg $l^{-1}$ NAA and 1 mg $l^{-1}$ kin resulted both in the highest frequency of callus induction and green plant regeneration for interspecific hybrids between *O. sativa* cv. Savitri and *O. rufipogon*, a presumed progenitor of *O. sativa* (Rout & Sarma, 1991).

Efficient production of microspore callus with high morphogenic potential requires both auxin and cytokinin in the medium (Huang *et al.*, 1986; Kavi Kishore *et al.*, 1989). Callus formation and plant regeneration could be enhanced by 1 mg $l^{-1}$ kin and 2 mg $l^{-1}$ NAA, but 2 mg $l^{-1}$ kin resulted in more albino plants. Kavi Kishore *et al.* (1989) regenerated plants at a higher frequency on $R_2$ medium containing 2 mg $l^{-1}$ each kin and IAA or LS medium with 1 mg $l^{-1}$ IAA and 4 mg $l^{-1}$ kin. Similarly, Draz *et al.* (1991) observed plant regeneration on media with 5 mg $l^{-1}$ kin and 1 mg $l^{-1}$ IAA or 2 mg $l^{-1}$ kin and 1 mg $l^{-1}$ IAA. Anther-derived rice callus could be placed into long-term maintenance for *in vitro* selection schemes by incubation on MS medium with 2% sucrose, 3% sorbitol, and 4.5 μM 2,4–D (KrishnaRaj & SreeRangasamy, 1993).

*β*-ecdysone, an insect hormone from *Cyanotis arachnoidea* C.B. Clarke, at 1–2 mg $l^{-1}$, enhanced anther callusing in wheat and rice (Hu *et al.*, 1978) but its effect was different from that of 2,4–D and kin because it promoted both differentiation of pollen embryos and callus growth (He *et al.*, 1980).

## 10. Effect of organic substances

Addition of organic substances, viz. CW, YE, potato extract, CH, lactalbumin hydrolysate, etc., in the medium has promoted callus formation from rice anthers (Guha *et al.*, 1970, Guha-Mukherjee, 1973; Wang *et al.*, 1974; Oono, 1975). By contrast, several studies have shown equally high frequency of callus induction and plant regeneration on media devoid of these substances (Chen, 1977; Chaleff & Stolarz, 1981; Lee & Chen, 1987; Chen *et al.*, 1982; Tsay *et al.*, 1982). Zapata *et al.* (1986a) reported an inhibitory effect of mashed banana whereas addition of CW (10%) increased the number of calli 1.4 times greater than the control. Hirabayashi *et al.* (1992) observed that the addition of aspartic acid, glutamine, glutamic acid, tryptophan and CH increased the efficiency of callus formation and plant regeneration. Anther browning was reduced, suggesting a drop in ethylene-promoted senescence. Normally, anther browning develops as a result of quinone production leading to degeneration of wall tissue. This can be disadvantageous, as anther wall tissue helps to supply nutrients to the developing microspores. To overcome this problem, ginseng powder at 500–1000 mg $l^{-1}$ can be added to the medium to delay anther browning, resulting in increased callus induction frequency (Chen & Tsay, 1986).

## 11. Sucrose

Sucrose is essential for androgenic plant development and it functions both as a nutrient and an osmotic factor. It influences callus induction and green plant regeneration. Chen (1978) reported that the rate of callus formation and subsequent organogenesis increased as the sucrose concentration increased from 3 to 9%, but that calli initiated on a medium with 9% sucrose regenerated more albino plantlets. A combination of 6% sucrose in the callus induction medium and 3% in the plant regeneration medium gave the highest frequency of regenerated green plants. However, it has also been suggested that optimal sucrose concentration for callus induction is 3% (Kavi Kishor *et al.*, 1989), 4–5% (Huang *et al.*, 1978; Chu, 1981; Chaleff & Stolarz, 1981), 5% (Reddy *et al.*, 1985; Chen, 1978), or 6% (Chaleff, 1980). It seems that high sucrose concentration alters the osmotic potential of the medium without providing any additional nutritional requirement such that mannitol (50 mM) can be substituted for additional sucrose (Chaleff & Stolarz, 1981). We have used a combination of tryptophan (25 mg $l^{-1}$), mannitol (1% w/v) and sucrose (3% w/v) in agar-solidified medium and cold pretreatment (8°C; 7 days) of panicles (collected from the primary tillers at the commencement of flowering) to enhance callusing from cultured anthers (Fig. 4). Up to 40% of the anther-derived calli regenerated green and albino plants (in a 1:1 ratio).

## 12. Shock pretreatment

The normal gametophytic pathway of pollen can be altered to sporophytic pathways by subjecting anthers or buds to various stresses, especially cold or heat shock treatments.

### 12.1. *Cold treatment*

Nitsch & Norreel (1973) found that the yield of pollen embryos/plants increased remarkably in *Datura innoxia* after flower buds were subjected to low temperature treatment for a short period before culturing. The beneficial effects of cold/hot pretreatments to anthers/panicles/buds have been demonstrated in anther culture of many species (Bajaj, 1990). With rice, a mild temperature shock of up to 11 days enhanced the frequency of callusing from 12.5 to 32.8% whereas a longer cold treatment of 25 days, although still enhancing callus formation, resulted in a higher frequency of albinos among the regenerants (Gupta & Borthakur, 1987). Cold treatment also delays anther senescence thus allowing sufficient time for anther wall tissue to nurture developing microspores (Zhang & Chu, 1986). In cold treated microspores, androgenic division began from the first day of culture, whereas in untreated microspores, little cell division occurred (Tian & Chen, 1983a). The beneficial effects of cold treatment have also been reported by Chaleff *et al.* (1975); Chen *et al.* (1981); Zapata *et al.* (1982); Lai & Chen (1984); Tsay *et al.* (1988); Ogawa *et al.* (1992). For optimal anther culture response, the cold pretreatment requirement varied depending on the cultivar (Yamaguchi *et al.*, 1990).

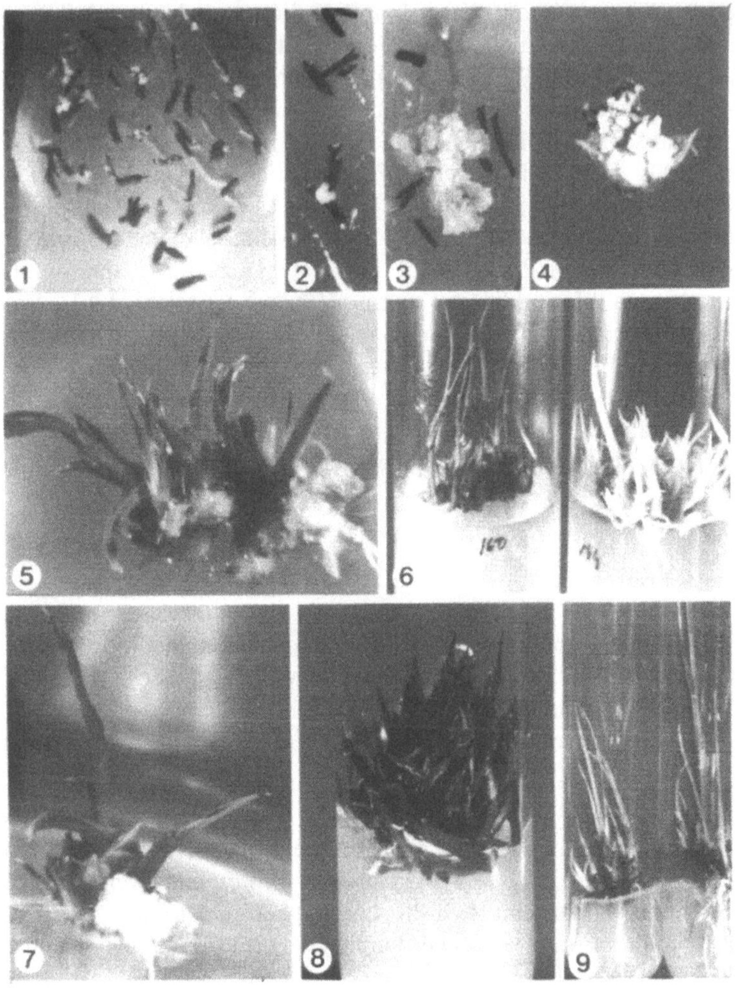

*Figure 4.* Anther culture and plant regeneration in Indica rice. (1–4) Callus initiation and its subsequent proliferation from cultured anthers on $N_6$ medium. (5–8) Regeneration of green and albino shoots fron anther-derived calli subcultured on MS + kin (0.5 mg $l^{-1}$). (9) Shoot elongation and root induction on half strength MS medium.

### 12.2. *Heat shock treatment*

A short heat treatment has been beneficial for callus induction from anthers. Reddy *et al.* (1985) induced callus from rice anthers by pretreating them at 35°C for 5 min followed by cold treatment at 10°C for seven days. Similarly,

Zapata & Torrizo (1986) reported an increase in callus induction frequency from anthers treated at 35°C for 15 min prior to cold treatment at 10°C for seven days. Sathish *et al.* (1995) found that a heat pretreatment of immersing the panicles in a water bath at 32°C for 2.5 or 5 h enhanced the ability of anthers to initiate callus.

## 13. Mode of action of shock pretreatment

The mode of action of shock pretreatments is not well understood. Nitsch & Norreel (1973) and Nitsch (1974) hypothesized that low temperature pretreatment resulted in the formation of two equal-sized nuclei and such microspores were more embryogenic than those with vegetative and generative nuclei. Chen *et al.* (1991) proposed that the action of pretreatment may be either to shut down or inhibit the function of genes responsible for gametophytic development. Microspores that have been affected by the pretreatment would be comparatively less differentiated and therefore, more easily shifted to sporophytic development. Vergne *et al.* (1993) found accumulation of a 32 kDa protein (MAR 32) in the anthers of maize tassels of responsive genotypes after cold treatment. A positive correlation was observed between the rate of embryo formation and the level of this protein.

## 14. Microspore isolation and culture

Microspore culture has several advantages over anther culture because microspores are haploid single cells that can readily be genetically manipulated (Dunwell, 1985). Moreover, microspore culture enables developing microspores to form callus. The first report on the culture of isolated mature pollen culture appeared in *Brassica oleracea* and *B. oleracea* × *B. alboglabra* (Kameya & Hinata, 1970). Ku & Huang (1973) were the first to obtain callus from rice pollen isolated from anthers. There are two methods of microspore isolation: a) naturally shed microspores in the culture medium after preculture of anthers, and b) mechanical means (by crushing or magnetic stirring). Chen *et al.* (1981) and Wang *et al.* (1979) regenerated plants from isolated pollen calli by giving panicles a cold treatment at 8–10°C following which the dehisced pollen was cultured. The naturally shed pollen yielded more calli and plantlets than mechanically isolated rice pollen cultures (Chen *et al.*, 1980, 1981). Similar results have been obtained in tobacco (Sunderland & Roberts, 1977) and barley (Sunderland & Xu, 1982). Cold shock of panicles at 10°C for 10–15 days before microspore isolation was beneficial (Chen, 1983). Pre-culture of anthers was important because direct culture resulted in only a few microspores undergoing division (Chen, 1986a). Ficoll (3–20%) has used to isolate microspore fractions: the highest frequency of callus formation was observed from the microspore fraction present in the gradient layer of 12% Ficoll (Zuo *et al.*, 1983). Datta *et al.* (1990a) obtained pollen plants by supplementing medium with 10% Ficoll.

The rate of success in rice microspore culture has been poor due to the low frequency of green plant regeneration (Chen, 1986a). Jia *et al.* (1987) regenerated plants from microspore cultures of several Japonica rice cultivars. Cho & Zapata (1988) found that large microspores (50–58 μm diam) with thin, pink-coloured cell walls produced embryos, whereas division of 40 μm microspores with thick cell walls was not observed. Glutamine and proline at 1 mM concentration had a stimulatory effect on microspore cultures. Plant regeneration from microspore-derived calli of Indica rice was genotype dependent (Cho & Zapata, 1990). Nitsch & Godard (1978) found that endogenous proline and glutamine first increased, then decreased, during growth of pollen embryos in maize and millet which implied that there was both production as well as utilization of these two amino acids that had been found to have a stimulatory effect on microspore culture of rice. Datta *et al.* (1990a,b) obtained plants from microspore cultures of both Japonica and Indica rice cultivars and also reported fertile plant regeneration from protoplasts isolated from microspore-derived cell suspensions of Indica rice. These protoplasts were amenable to genetic engineering (Datta *et al.*, 1990c). Ogawa *et al.* (1992) reported that more than 20 day cold treatment (10°C) was required for microspores of Indica rice to divide; the highest colony formation (0.4%) was obtained after a cold treatment at 10°C for 25 days. A three day sugar starvation followed by incubation in R2 salts with 1 mM $KH_2PO_4$ resulted in the highest frequency of callus formation of a Japonica rice cultivar (Ogawa *et al.*, 1994). Despite several reports on microspore culture, anther culture is still more widely used for producing haploids/doubled haploids of rice. Concerted efforts are needed to develop protocols that enhance the rate of green plant regeneration from microspore cultures.

## 15. Ploidy of anther-derived plants

Anther-derived rice plants comprise a mixed population with ploidy ranging from 1*x* to 5*x*, although most are haploid (30–35%) or diploid (50–65%) (Chen *et al.*, 1983; Mercy & Zapata, 1986; Rout & Sarma, 1990). Many factors, such as developmental stage of microspores at the time of culture, nature of pretreatment, genotype of the donor plants, composition of the culture medium, and age of the cultivars, affect the ploidy of the regenerated plants. Both endoreduplication and nuclear fusion have been demonstrated to occur during rice anther culture and are presumed to result in diploid regenerants (Chen & Wu, 1983). The frequency of doubled haploid regenerants has been enhanced 1.5–2.5-fold by the incorporation of 250 or 500 mg $l^{-1}$ colchicine in the induction medium (Alemanno & Guiderdoni, 1994). By anther culture of trisomic rice plants, both tetrasomics and $n + 1$ plants have been regenerated (Wang & Iwata, 1991).

Haploid and diploid rice exhibit distinct morphological differences. Haploids are characterized by the absence of ligules and auricles, approximately twice as many tillers, and 60–70% of plant height, panicle length and leaf length compared to diploids. However, glume length has been demonstrated to be the most diagnostic character for distinguishing the ploidy of anther-derived rice (Nakamura *et al.*, 1994). For breeding purposes, haploid plants must be diploidized for restoration of fertility. Chromosome doubling has been done by submerging the base of haploid seedlings in 0.1–0.2% colchicine solution for 12–24 h (Chung *et al.*, 1991).

## 16. Variation and inheritance of traits in regenerated plants

Anther-derived plants are identified as the first generation of pollen plants ($H_1$). Their immediate progeny are known as $H_2$. The genotype may not be well-expressed in the $H_1$ generation but there is no expected genotypic difference between an $H_1$ plant and its $H_2$ generation. Segregation among the anther-derived progeny of $F_1$ hybrids includes plants that resemble the male and female parents of the hybrids, intermediate forms, and new types. Some gametic selection has been demonstrated after anther culture of Japonica × Indica hybrids but this was not considered to be a major limitation to the implementation of anther culture in rice breeding (Guiderdoni, 1991). Most traits are uniform among plants derived from DHs (Schaeffer, 1982), resulting in small coefficients of variation for quantitative traits, even after 4–5 generations. No significant differences for traits, like number of effective tillers, 1000 grain weight, plant height, and panicle length have been observed in the progenies of pollen plants over several generations (Zhang, 1982). Sometimes, a few $H_2$ lines have exhibited segregation for certain traits that may be attributed to the occurrence of gametoclonal variation, origin of plants from somatic tissue instead of pollen, and outcrossing as a result of male sterility (Zhang *et al.*, 1992).

Field studies have been conducted for several anther-derived rice populations with specific traits under evaluation. Zapata *et al.* (1991a) evaluated the saline tolerance of anther-derived lines from hybrids between salt tolerant and agronomic selections of Japonica cultivars; some of the anther-derived line were equal to the high-yielding parent under non-stressed conditions but were superior under salt stress conditions. N'Guessan *et al.* (1994) have identified seven promising anther culture lines among 43 tested that exhibited resistance to the rice water weevil (*Lissorhoptrus oryzophilus* Kuschel). Plants with larger seeds, higher levels of seed protein, shorter stature, and more highly tillered were recovered among the selfed anther-derived plants of cv. Calrose 76 (Schaeffer *et al.*, 1984). The anther-derived plants from an interspecific *O. sativa* × *O. rufipogon* exhibited transgressive segregation, some of which were phenotypically close to *O. sativa* but with a few traits of the wild parent.

## 17. Ovary culture

Induction of haploids from megaspores was first reported by Uchimiya *et al.* (1971) by culturing unpollinated ovaries of *Zea mays* and ovules of *Solanum melongena*. They observed division of haploid cells in callus tissue and postulated that *in vitro* induction of haploid plants from the angiosperm megagametophyte was possible. Further progress on haploid plant production from ovary culture has been made by many researchers (Asselin de Beauville, 1980; San Noeum & Ahmadi, 1980; Kuo, 1981; Zhou & Yang, 1981a,b,c; Yang & Zhou, 1982). In rice, unhusked flowers with pistil and stamens attached to the receptacle were highly responsive in liquid medium; however, cultures were less responsive when stamens were removed, and the worst response was obtained when isolated pistils were used (Zhou & Yang, 1981a,c). Zhou & Yang (1981b) observed genotypic differences for production of gynogenic calli. The success of ovary culture is dependent on the developmental stage of the ovary. Although success has been reported with ovaries ranging from uninucleate to mature embryo sac stages (Zhou & Yang, 1981a; Kuo, 1981; Huang *et al.*, 1982), nearly mature embryo sacs have been preferred (San Noeum, 1976, 1979; Asselin de Beauville, 1980; Wang & Kuang, 1981).

The regulation of growth regulators to enhance gynogenesis but inhibit proliferation of somatic tissues has been critical for ovary culture (Cagnet-Sitbon, 1980; Huang *et al.*, 1982; Kuo, 1981; Zhou *et al.*, 1983). Zhou & Yang (1981b,c) studied the growth regulator requirement for gynogenesis and reported that, in the absence of MCPA (2-methyl,4-chlorophenoxyacetic acid), ovaries failed to enlarge and produced gynogenic calli. An increase in MCPA concentration from 0.125 to 8.0 mg $l^{-1}$ favoured ovary swelling but higher levels of auxin (more than 2 mg $l^{-1}$) stimulated callus formation from the ovary wall rather than from the embryo sac. A sucrose concentration of 3–6% was suitable for rice ovary culture whereas 9% was unsuitable (Zhou *et al.*, 1983). Compared to anther culture, ovary culture is inefficient because there is only one embryo sac per ovary as opposed to thousands of microspores per anther. However, the rate of induced embryo sacs has generally been higher than for microspores and the frequency of regenerated green plants has also been higher than for androgenesis. More effort is needed to develop this system to regenerate a higher frequency of green haploid plants.

## 18. Conclusion

Anther culture protocols for Japonica rice are now well-established and routinely used for developing homozygous DHs from crosses between elite cultivars. However, in Indica rice, despite several reports, anther culture success is limited, due to low frequency of callus induction and subsequent regeneration of green plants. Concerted efforts are required to enhance the frequency of responding microspores, to improve callus growth and subsequent green plant regeneration. The use of nurse cultures with maltose and proline has been reported to induce and improve the protoplast-to-plant system in recalcitrant Indica rices (Datta *et al.*, 1992; Ghosh-Biswas *et al.*, 1993; Jain *et al.*, 1995). Likewise, simple dehydration treatments have been found to promote plantlet regeneration from scutellum-derived calli (Tsukahara & Hirosawa, 1992). The extension of these techniques to anther culture may help in achieving greater success in a wide range of rice germplasm.

## 19. Acknowledgements

The authors wish to acknowledge a grant from The Rockefeller Foundation, New York, USA, for supporting the rice biotechnology research at Punjab Agricultural University.

## 20. References

Abe, K. (1992) Genealogical study on callus formation ability in anther culture of rice variety, Koshohikara. Jpn. J. Breed. 42: 403–413.

Aldemita, R.R. & F.J. Zapata (1991) Anther culture of rice: effects of radiation and media components on callus induction and plant regeneration. Cereal Res. Commun. 19: 9–32.

Alemanno, L. & E. Guiderdoni (1994) Increased doubled haploid plant regeneration from rice (*Oryza sativa* L.) anthers cultured on colchicine-supplemented media. Plant Cell Rep. 13: 432–436.

Aruna, M. & G.M. Reddy (1988) Genotypic differences in callus initiation and plant regeneration from anthers of Indica rice. Curr. Sci. 57: 1014–1017.

Asselin de Beauville, M. (1980) Obtention d'haploïdes *in vitro* à partir d'ovaires non fecondés de riz, *Oryza sativa* L. C.R. Hebdomadaires Sci. Acad. Sci. 290D: 489–492.

Baenziger, P.S. & G.W. Schaeffer (1983) Dihaploids via anthers cultured *in vitro*. In: L.D. Owens (Ed.), Genetic Engineering: Applications to Agriculture. Beltsville Symp. Agric. Res. Vol. VII, pp. 269–284. Rowan and Allanheld, Totowa NJ.

Bajaj, Y.P.S. (1990) Biotechnology in Agriculture and Forestry, Vol. 12: Haploids in Crop Improvement I. Springer-Verlag, Berlin.

Beaumont, V. & B. Courtois (1990a) Comportement en androgenèse d'anthères de riz provenant de plantes traitées avec un agent chimique d'hybridation. Agron. Trop. 45: 95–100.

Beaumont, V. & B. Courtois (1990b) Anther culturability of rice plants treated with male gametocide chemicals. Intl. Rice Res. Newsl. 15: 9.

Blaydes, D.F. (1966) Interaction of kinetin and various inhibitors in the growth of soybean tissues. Physiol. Plant. 18: 748–753.

Cagnet-Sitbon, M. (1980) Recherches préliminaires sur la production d'haploïdes de *Gerbera* par culture d'anthères et d'ovules non fecondées *in vitro*. Thèse de 3eme cycle Amélior. Plantes. Université Paris-Sud, Centre d'Orsay, Paris.
Chaleff, R.S. (1978) Anther culture as a rice breeding technique. Intl. Rice Res. Newsl. 3: 2–3.
Chaleff, R.S. (1980) Innovative approaches to rice breeding. In: Tissue Culture in Rice Improvement: An Overview, pp. 81–91. IRRI, Manila.
Chaleff, R.S. & A. Stolarz (1981) Factors influencing the frequency of callus among cultured rice *Oryza sativa* anthers. Physiol. Plant. 51: 201–206.
Chaleff, R.S. & A. Stolarz (1982) The development of anther culture as a system for *in vitro* mutant selection. In: Rice Tissue Culture Planning Conference, pp. 63–74. IRRI, Los Banos.
Chaleff, R.S., S.R. Hill & J.M. Dunwell (1975) Rice anther culture. In: Annual Report John Innes Institute, Norwich, UK, pp. 64–66.
Chen, C.C. (1976) Studies on the anther culture of rice: Pollen stage and low temperature treatment. Natl. Sci. Counc. Monthly (Taipei) 4(2): 2187–2190.
Chen, C.C. (1977) *In vitro* development of plants from microspores of rice. In Vitro 13: 484–489.
Chen, C.C. (1978) Effects of sucrose concentration on plant production in anther culture of rice. Crop Sci. 18: 905–906.
Chen, C.C. & M.H. Lin (1976) Induction of rice plantlets from anther culture. Bot. Bull. Acad. Sin. (Taipei) 17: 18–23.
Chen, C.C. & C.M. Chen (1979) A method for anther culture of rice. Tissue Cult. Assoc. Manual 5: 1051–1053.
Chen, C.C. & Y.H. Wu (1983) Segmentations in microspores of rice during anther culture. Proc. Natl. Sci. Counc. 7B: 151–157.
Chen, C.C., W.L. Chiu, L.J. Yu, S.S. Ren & W.J. Yu (1983) Genetic analysis of anther-derived plants of rice: independent assortment of unlinked genes. Can. J. Genet. Cytol. 25: 324–327.
Chen, C.C., H.S. Tsay & C.R. Huang (1991) Factors affecting androgenesis in rice *Oryza sativa* L. In: Y.P.S. Bajaj (Ed.), Biotechnology in Agriculture and Forestry, Vol. 14, pp. 193–215. Springer-Verlag, Berlin.
Chen, C.M. & C.C. Chen (1979) Selection and regeneration of 5–methyltryptophan resistant rice plant from pollen callus. Natl. Sci. Counc. Monthly (Taipei) 7: 199–203.
Chen, C.M., C.C. Chen & M.H. Lin (1982) Genetic analysis of anther-derived plants of rice. J. Hered. 73: 49–52.
Chen, J.J. & H.S. Tsay (1984) The callus forming ability of rice anther derived from tillers of different orders and from spikelets of different positions and branches in the panicle. J. Agric. Res. China 33: 354–362.
Chen, J.J. & H.S. Tsay (1986) Effects of ginseng on rice anther culture. J. Agric. Res. China 35: 139–144.
Chen, J.J., J.Y. Hsu & H.S. Tsay (1986) Effect of iron on rice anther culture. J. Agric. Res. China 35: 244–252.
Chen, L.Z., B.Z. Lai, Q.H. Liaoo & X.S. Cai (1982) Medium evaluation for rice anther culture. J. Agric. Res. China 31: 283–290.
Chen, Y. (1983) Anther and pollen culture of rice in China. pp. 11–26. In: Proc. of workshop on Cell and Tissue Culture Techniques for Cereal Crop Improvement, Science Press Beijing and Int. Rice Res. Institute, Manila, Philippines.
Chen, Y. (1986a) Anther and pollen culture of rice. In: H. Hu and H. Yang (Eds.), Haploids of Higher Plants *in Vitro*, pp. 3–25. China Academic Publishers, Beijing/Springer-Verlag, Berlin.
Chen, Y. (1986b) The inheritance of rice pollen plant and its application in crop improvement. In: H. Hu and H. Yang (Eds.), Haploids of Higher Plants *in Vitro*, pp. 118–136. China Academic Publishers, Beijing/Springer-Verlag, Berlin.
Chen, Y., R.F. Wang, W.Z. Tian, Q.X. Zuo, S.W. Zheng, D.Y. Lu & G.H. Zhang (1980)

Studies on pollen culture *in vitro* and induction of plantlets in *Oryza sativa* sub. sp. Keng. Acta Genet. Sin. 7: 46–54.

Chen, Y., Q.X. Zuo, S.Y. Li, D.Y. Lu & S.W. Zheng (1981) Study of factors associated with the induction of green plants from isolated rice pollen grains *in vitro*. Acta Genet. Sin. 8: 158–163.

Chen, Z. & Q. Chen (1993) Genetic studies of rice (*Oryza sativa* L.) anther culture response. Plant Cell Tiss. Organ Cult. 34: 177–182.

Cheng, J.C. (1978) A preliminary study on organogenesis of pollen callus of rice. In: Proc. Symp. Anther Cult. Science Press, Peking, pp. 126–132.

Chengmei, Z. & Z. Zhengua (1990) Effect of *Luffa cylindrica* Roem. exudate on plant induction from rice anthers. Intl. Rice Res. Newsl. 15: 6–7.

Cho, M.S. & F.J. Zapata (1988) Callus formation and plant regeneration in isolated pollen culture of rice *Oryza sativa* L. cv. Taipai 309. Plant Sci. 58: 239–244.

Cho, M.S. & F.J. Zapata (1990) Plant regeneration from isolated microspore of Indica rice. Plant Cell Physiol. 31: 881–885.

Chu, C.C. (1981) The $N_6$ medium and its applications to anther culture of cerael crops. In: Proc Symp. Plant Tissue Culture, pp. 43–50. Pitman Press, Boston.

Chu, C.C. (1982) Anther culture of rice and its significance in distant hybridization. In: Rice Tissue Culture Planning Conference, pp. 47–53. IRRI, Los Banos.

Chu, C.C., C.S. Wang, C.S. Sun, C. Hsu, K.C. Yin, C.Y. Chu & F.Y. Bi (1975) Establishment of an efficient medium for anther culture of rice through comparative experiments in the nitrogen sources. Sci. Sin. 18: 659–668.

Chung, G.S., S.J. Yang & B.G. Oh (1991) Reliable simple method for producing rice plants through anther culture. In: Rice Genetics II. Proc. Second Intl. Rice Genet. Symp. 14–18, May 1990, pp. 791–792.

Courtois, B. (1992) Influence of type of vessel on regeneration of rice anther calli. Intl. Rice Res. Newsl. 17: 6.

Courtois, B. & J. Taillebois (1990) Comportement en androgenèse de lignées de riz porteuses d'une stérilité mâle d'origine genocytoplasmique. Agron. Trop. 45: 101–105.

Datta, S.K., K. Datta & I. Potrykus (1990a) Embryogenesis and plant regeneration from microspores of both "Indica" and "Japoncia" rice *Oryza sativa*. Plant Sci. 67: 83–88.

Datta, S.K., K. Datta & I. Potrykus (1990b) Fertile Indica rice plants regenerated from protoplasts isolated from microspore derived cell suspensions. Plant Cell. Rep. 9: 253–256.

Datta, S.K., A. Peterhans, K. Datta & I. Potrykus (1990c) Genetically engineered fertile Indica-rice recovered from protoplasts. Bio/Technology 8: 736–740.

Datta, K., I. Potrykus & S.K. Datta (1992) Efficient fertile plant regeneration from protoplasts of the Indica rice breeding line IR72 (*Oryza sativa* L.). Plant Cell Rep. 11: 229–233.

de Beauville, A.M. (1980) Obtention d'haploides *in vitro* à partir ovaires non-jecondés de Riz *Oryza sativa* L.C.R. Acad. Sci. Paris 296D: 489–492.

Draz, A.E., F.J. Zapata & G.S. Khush (1991) Interaction of media for callus induction and for plant regeneration in rice anther culture. Intl. Rice Res. Newsl. 165: 7–8.

Dunwell, J.M. (1985) Anther and ovary culture. In: S.W.J. Bright & M.G.K. Jones (Eds.), Cereal Tissue and Cell Culture, pp. 1–44. Kluwer Academic Publishers, Dordrecht.

Gamborg, O.L., R.A. Miller & K. Ojima (1968) Nutrient requirements of suspension cultures of soybean root cells. Exptl. Cell Res. 50: 151–158.

Genovesi, A.D. & C.W. Magill (1979) Improved rate of callus and green plant production from rice anther culture following cold shock. Crop Sci. 19: 662–664.

Ghosh-Biswas, C.G. & F.J. Zapata (1993) High-frequency plant regeneration from protoplasts of Indica rice (*Oryza sativa* L.) using maltose. J. Plant Physiol. 141: 470–475.

Glaszmann, J.C. (1987) Isozymes and classification of Asiatic rice varieties. Theor. Appl. Genet. 74: 21–30.

Gosal, S.S., A.S. Sindhu, G.S. Khehra & D.S. Brar (1991) Somatic cell, anther culture and plant regeneration in rice. Proc. Symposium on Recent Advances in Rice Genetics and Tissue Culture, Osmania University, Hyderabad, India.

Gosal, S.S., H.S. Dhaliwal, J.S. Sandhu, M.S. Khinda & P.K. Agrawal (1993) Somatic cell, anther and protoplast culture in Indica rices. Presented at the Annual Meeting of Rockefeller Foundation held on Feb. 2–5 at Chiang Mai, Thailand.

Guha, S. & S.C. Maheshwari (1964) *In vitro* production of embryos from anthers of *Datura.* Nature 204: 497.

Guha, S. & S.C. Maheshwari (1966) Cell division and differentiation of embryos in the pollen grains of *Datura in vitro.* Nature 212: 97–98.

Guha, S., R.D. Iyer, N. Gupta & M.S. Swaminathan (1970) Totipotency of gametic cells and the production of haploids in rice. Curr. Sci. 39: 174–176.

Guha-Mukherjee, S. (1973) Genotypic differences in the *in vitro* formation of embryoids from rice pollen. J. Exp. Bot. 24: 139–144.

Guiderdoni, E. (1991) Gametic selection in anther culture of rice (*Oryza sativa* L.). Theor. Appl. Genet. 81: 406–412.

Guiderdoni, E., E. Galinato, J. Luistro & G. Vergara (1992) Anther culture of tropical Japonica × Indica hybrids of rice (*Oryza sativa*) L. Euphytica 62: 219–224.

Gupta, H.S. & D.N. Borthakur (1987) Improved rate of callus induction from rice anther culture following microscopic staging of microspores in iron alum-haematoxylin. Theor. Appl. Genet. 74: 95–99.

Hadi, S. & S.K. Raina (1989) Variation in anther culture efficiency among donor plant tillers in rice. Curr. Sci. 58: 1397–1398.

Harn, C. (1969) Studies on the anther culture of rice. Korean J. Breed. 1: 1–11.

Heszky, L.E., L.S. Nam, I.K. Simon, E. Kiss, K. Lokos & D. Birth (1991) *In vitro* studies on rice in Hungary. In: Y.P.S. Bajaj (Ed.), Biotechnology in Agriculture and Forestry, Vol. 14, pp. 619–641. Springer-Verlag, Berlin.

He, J.B., A. Hu & L.P. Peng (1980) The influences of ß-ecdysone on the induction of wheat pollen plantlets in anther culture. Acta Phytophysiol. Sin. 6: 369–375.

Hirabayashi, T., S. Misoo, O.I. Kamijima & M. Sawano (1992) Effects of various amino acids and casein hydrolysate on anther culture of rice (*Oryza sativa* L.) Science Reports of Faculty of Agriculture, Kobe Univ. Japan 30: 23–29.

Hu, C., C.P. Ho & L.P. Peng (1978) Effect of $\beta$-ecdysome on induction of pollen plants in rice and wheat In: Proc. Symp Anther Cult. Science Press, Peking, p. 270.

Hu, C., S.C. Huang, C.P. Ho, H.C. Liang, C.C. Chuang & L.P. Peng (1981) On the inductive conditions of rice pollen plantlets in anther culture. In: Proc. Symp. Plant Tissue Cult., pp. 87–95. Pitman Press, Boston.

Hu, H. (1981) Advances in anther culture investigations in China. In: Proc. Symp. Plant Tissue Cult., pp. 3–10. Pitman Press, Boston.

Hu, H. (1985) Use of haploids for crop improvement in China. Genet. Manipulations Crops Newsl. CI: 11–23.

Hu, H. & J.Z. Zeng (1983) Development of new varieties via anther culture. In: P.V. Ammirato, D.A. Evans, W.R. Sharp & Y. Yamada (Eds.), Handbook of Plant Cell Culture, Vol. 3, pp. 65–90. Macmillan Publishing Co., New York.

Huang, H.S., T.H. Ling, P.I. Tseng, Y.L. Shien & P. Shi (1978) Studies on medium component in anther culture of *Oryza sativa* sub sp. Hsein by mathematical methods. In: Proc. Symp. Plant Tissue Culture, Science Press, Peking pp. 244–246.

Huang, Q.F., H.Y. Yang & C. Zhou (1982) Embryological observation on ovary culture of unpollinated young flowers in *Hordeum vulgare* L. Acta Bot. Sin. 24: 295–300.

Huang, De-li, C.S. Zhao & Z.S. Zhao (1983) The effect of temperature on the frequency of albino plantlets in rice anther culture. In: Shen Jinhua *et al.* (Eds.), Studies on Anther Cultured Breeding in Rice, pp. 106–109. Agricultural Press, Beijing.

Huang, C.R., Y.N. Wu & C.C. Chen (1986) Effect of plant growth substances on callus formation and plant regeneration in anther culture of rice. In: Rice Genetics: Proc. Intl. Rice Genetics Symp., pp. 763–771. IRRI, Manila.

Inoue, M. & E. Maeda (1981) Stimulation of shoot bud and plantlet formation in rice callus cultures by two step culture method using abscisic acid and kinetin. Jpn. J. Crop Sci. 50: 318–322.

IRRI, 1989. Towards 2000 and Beyond. Intl. Rice. Res. Inst., Manila.

Iyer, R.D. & S.K. Raina (1972) The early ontogeny of embryoids and callus from pollen and subsequent organogenesis in anther culture of *Datura metel* and rice. Planta 104: 146–156.

Jain, R.K., G.S. Khehra, S.H. Lee, N.W. Blackhall, R. Merchant, M.R. Davey, J.B. Power, E.C. Cocking & S.S. Gosal (1995) An improved procedure for plant regeneration from Indica and Japonica rice protoplasts. Plant Cell Rep. 14: 515–519.

Jia, Y.J., T. Abe & Y. Futsuhara (1987). Int. Rice Genet. Newsl. 4: 109–110.

Kameya, T. & K. Hinata (1970) Induction of haploid plants from pollen grains of *Brassica*. Jpn. J. Breed. 20: 82–87.

Karim, N.H. & F.J. Zapata (1990) One step rice plantlet development through anther culture. Indian J. Plant Physiol. 33: 119–124.

Kavi Kishor, P.B., M. Aruna & G.M. Reddy (1989) Plant regeneration from haploid callus of Indica rice. Proc. Indian Natl. Sci. Acad. 55B: 193–202.

Kim, H.S., Y.T. Lee, S.Y. Lee & T.S. Kim (1991) Anther culture efficiency in different rice genotypes under different cold pretreatment durations and culture temperatures. Research Reports of the Rural Development Administration. Bio/Technology 33: 5–13.

KrishnaRaj, S. & S.R. SreeRangasamy (1993) *In vitro* salt tolerance screening in long-term anther cultures of rice (*Oryza sativa* L.) variety IR 50. J. Plant Physiol. 142: 754–758.

Ku, M.K. & T.N. Huang (1973) Callus formation and their cloning from isolated pollen of rice and wheat. Genet. Commun. 1: 28–31.

Kuo, C.S. (1981) The preliminary studies on *in vitro* culture of unfertilized ovaries of rice. Acta Bot. Sin. 24: 33–38.

Lai, M. & L. Chen (1984) Induction of rice plantlets from anther culture. Bot. Bull. Acad. Sin. 17: 18–24.

Lee, F.M. & C.C. Chen (1987) Nuclear fusion in cultured microspores of barley. Plant Cell Rep. 6: 191–193.

Lee, J. & C.C. Chen (1982) Genetic and histological evidence for microspore origin of anther-derived plants of rice. Taiwania 27: 86–92.

Lenka, N. & G.M. Reddy (1994) Role of media, plant growth regulators in callusing and plant regeneration from anthers of Indica rice. Proc. Indian Natl. Sci. Acad. 60B: 87–92.

Ling, D.H., M.F. Chen, Z.R. Ma, C.Y. Liang & B.Y. Chen (1990) Anther culture of photoperiod sensitive genic male sterile rice. Acta Bot. Sin. 6: 152–158.

Linsmaier, E.M. & F. Skoog (1965) Organic growth factor requirements of tobacco tissue culture. Physiol. Plant. 18: 100–127.

Marassi, M.A., O.A. Bovo, G.L. Lavia & L.A. Mroginski (1993) Regeneration of rice double haploids using a one step culture procedure. J. Plant Physiol. 141: 610–614.

McKenzie, K.S., C.N. Bollich, J.N. Rutger & K.A. Kuenzel Moldenhauer (1987) Rice. In: W.R. Fehr (Ed.), Principles of Cultivar Development, Vol. 2: Crop Species, pp. 487–532. Macmillan Publishing Co., New York.

Meifang, L.I. (1992) Anther culture breeding of rice at the CAAS. In: K. Zheng and T. Murashige (eds.) Anther Culture for Rice Breeders. pp. 75–85. Hangzhou, China.

Mercy, S.T. & F.J. Zapata (1986) Chromosomal behavior of anther culture derived plants of rice. Plant Cell Rep. 5: 215–218.

Mercy, S.T. & F.J. Zapata (1987) Position of anthers at plating and its influence on anther callusing in rice. Plant Cell Rep. 6: 318–319.

Miah, M.A.A., E.D. Earle & G.S. Khush (1985) Inheritance of callus formation ability in anther cultures of rice, *Oryza sativa* L. Theor. Appl. Genet. 70: 113–116.

Nakamura K., H. Suzuki, K. Hattori & Y. Futsuhara (1994) Identification of ploidy level of the regenerated plants by anther culture in rice. Breeding Sci. 44: 19–22.

Narasimman, R. & S.R.S. Rangasamy (1993) Comparison of fertility between the $F_1$, $F_2$ and anther derived lines in the crosses of Indica Japonica and Japonica Indica in rice (*Oryza sativa* L.). Euphytica 66: 19–25.

N'Guessan, F.K., S.S. Quisenberry & T.P. Croughan (1994) Evaluation of rice anther culture lines for tolerance to the rice water weevil (Coleoptera: Curculionidae). Environ. Entomol. 23: 331–336.

Niizeki, H. & K. Oono (1968) Induction of haploid rice plant from anther culture. Proc. Jpn. Acad. 44: 554–557.

Nishi, T. & S. Mitsuoka (1969) Occurrence of various ploidy plants from anther and ovary culture of rice plant. Jpn. J. Genet. 44: 341–346.

Nitsch, C. (1974) Pollen culture – a new technique for mass production of haploid and homozygous plants. In: K. Kasha (Ed.), Haploids in Higher Plants: Advances and Potential, pp. 123–135. The University of Guelph, Guelph.

Nitsch, C. & B. Norreel, 1973. Effet d'un choc thermique sur le pouvoir embryogène du pollen de *Datura innoxia* cultivé dans l'anthère et isolé de l'anthère. C.R. Acad. Sci. Paris 276D: 303–306.

Nitsch, C. & M. Godard (1978) The role of hormones in promoting and developing growth to select new varieties in sterile culture. In: T.K. Scott (Ed.), Plant Regulation and World Agriculture, pp. 49–62. Plenum Press, New York.

Ogawa, T., T. Hagio & Y. Ohkawa (1992) Plant regeneration from isolated pollen grains in Indica type rice (*Oryza sativa* L.). Jpn. J. Breed. 42: 675–679.

Ogawa, T., H. Fukuoka & Y. Ohkawa (1994) Induction of cell division of isolated pollen grains by sugar starvation in rice. Breeding Sci. 44: 75–77.

Oka, H.I. (1988) Indica/Japonica differentiation of rice cultivars. In: Origin of Cultivated Rice, pp. 141–179. Japan Science Soc. Press, Tokyo/Elsevier, Amsterdam.

Oono, K. (1975) Haploid plants of rice (*Oryza sativa*) by anther culture and their use for breeding. Bull. Natl. Inst. Agric. Sci. 26D: 139–222.

Quimio, C.A. & F.J. Zapata (1990) Diallel analysis of callus induction and green plant regeneration in rice anther culture. Crop Sci. 30: 188–192.

Raina, S.K., P. Sathish & K.S. Sarma (1987) Plant regeneration from *in vitro* cultures of anthers and mature seeds of rice *Oryza sativa* L. cv. Basmati-370. Plant Cell Rep. 6: 43–45.

Raina, S.K., S.M. Balachandran, S.S. Virmani & F.J. Zapata (1989) Improved medium for efficient anther culture of some Indica rice haploids. Intl. Rice Res. Newsl. 14: 4.

Reddy, Y.S., S. Leelavathi & S.K. Sen (1985) Influence of genotype and culture medium on microspore callus, induction and green plant regeneration in anthers in *Oryza sativa*. Physiol. Plant. 63: 309–314.

Reiffers, I. & A.B. Freire (1990) Production of doubled haploid rice plants (*Oryza sativa*) L. by anther culture. Plant Cell Tiss. Organ Cult. 21: 165–170.

Rout, J.R. & N.P. Sarma (1990) Comparative morphological studies of microspore derived and $F_2$ plants obtained from an interspecific rice hybrid (*Oryza sativa* Linn. × *O. rufipogon* Griff.). Plant Breed. 105: 283–291.

Rout, J.R. & N.P. Sarma (1991) Anther callus induction and green plant regeneration at high frequencies from an interspecific rice hybrid *Oryza sativa* Linn. × *O. rufipogon* Griff. Euphytica 54: 155–159.

Rout, J.R., N.P. Sharma & G.J.N. Rao (1989) Effect of potato-2 medium on anther culture of interspecific rice hybrids. Ann. Bot. 63: 621–624.

Rush, M.C. & Q.Q. Shao (1982) Rice improvement through cell and tissue culture. In: Rice Research Strategies for the Future, pp. 159–183. Intl. Rice Res. Inst., Manila.

San Noeum, L.H. (1976) Haploïdes d'*Hordeum vulgare* L. par culture *in vitro* d'ovaires non fecondés. Ann. Amélior. Plant. 26: 751–754.

San Noeum, L.H. (1979) *In vitro* induction of gynogenesis in higher plants. In: Proc. Conf. Broadening Genetic Base Crops Wageningen 1978, pp. 327–329. Pudoc, Wageningen.

San Noeum, L.H. & N. Ahmadi (1980) Variability of doubled haploides from *in vitro* androgenesis and gynogenesis in *Hordeum vulgare* L. Colloque NSP, CNRS Orsay.

Sandhu, J.S., M.S. Gill & S.S. Gosal (1993a) Callus induction and plant regeneration from cultured anthers of Indica rice varieties. Plant Tissue Cult. 3: 17–21.

Sandhu, J.S., S.S. Gosal, G.S. Sidhu, H.S. Grewal & H.S. Dhaliwal (1993b) Anther culture of $F_2$ plants derived from high yielding × quality type rice. In: M.M. Verma (Ed.), Heterosis Breeding in Crop Plants: Theory and Application, pp. 114–115. Punjab Agric. Univ. Ludhiana, India.

Sanint, L., C.P. Martinez, A. Ramirez & Z. Lentini (1992) Rice anther culture versus conventional breeding: A cost/benefit analysis. CIAT Working Document.

Sathish, P., O.L. Gamborg & M.W. Nabors (1995) Rice anther culture: callus initiation and androclonal variation in progenies of regenerated plants. Plant Cell Rep. 14: 432–436.

Sato, Y.I., S. Chitrakon & H. Moushima (1986) The Indica-Japonica differentiation of native cultivars in Thailand and neighbouring countries. In: B. Napompeth & S. Subhadrabandhu (Eds.), New Frontiers in Breeding Researches, pp. 185–191. Kasetsari University, Bangkok.

Schaeffer, G.W. (1982) Recovery of heritable variability in anther-derived doubled-haploid rice. Crop Sci. 22: 1160–1164.

Schaeffer, G.W., F.T. Sharpe, Jr. & P.B. Cregan (1984) Variation for improved protein and yield from rice anther culture. Theor. Appl. Genet. 67: 383–389.

Shahjahan, A.K.M., N.H. Karim & S.A. Miah (1985) Culture conditions and callus forming ability of rice anthers. Intl. Rice Res. Newsl. 10: 22.

Shen, J., M. Li, Y. Chen & Z. Zhang (1983) Improving rice by anther culture. In: Cell and Tissue Culture Techniques for Cereal Crop Improvement, pp. 183–205. Science Press, Beijing.

Sohn, J.K., B.G. Oh, S.K. Lee & G.S. Chung (1987) Effects of liquid medium on callus induction and plant regeneration in rice anther culture. Research Reports of the Rural Development Administration – Crops – Korea Republic 29: 38–42.

Sun, C.S. (1981) Androgenesis of cereal crops. In: Proc. Symp. Plant Tissue Cult., pp. 117–123. Pitman Press, Boston.

Sun, Z.X., H.M. Si, X.Y. Zhan & S.H. Chen (1992) The effect of thermo-photoperiod for donor plant growth on anther culture of Indica rice. In: C.B. You (Ed.), Biotechnology in Agriculture. Proceedings of the First Asia-Pacific Conference on Agricultural Biotechnology, Beijing, China, 20–24 August 1992. Current Plant Science and Biotechnology in Agriculture 15: 361–364.

Sunderland, N. & J.M. Dunwell (1977) Anther and pollen culture. In: H.E. Street (Ed.), Plant Tissue and Cell Culture, pp. 223–265. Blackwell, Oxford.

Sunderland, N. & M. Roberts (1977) New approach to pollen culture. Nature (London) 270: 238.

Sunderland, N. & Z.H. Xu (1982) Shed pollen culture in *Hordeum vulgare*. J. Exp. Bot. 33: 1086–1095.

Tian, W.Z. & Y. Chen (1983a) Factors affecting induction and differentiation of callus in rice anther float culture. Acta Genet. Sin. 10: 362–368.

Tian, W.Z. & Y. Chen (1983b) Increasing differentiation frequency of green plantlets in rice anther float culture. In: Cell and Tissue Culture Techniques for Cereal Crop Improvement, p. 427. Science Press, Beijing.

Toriyama, K. & K. Hinata (1985) Panicle culture in liquid media for obtaining anther calli and protoplasts in rice. Jpn. J. Breed. 35: 449–452.

Torrizo, L.B. & F.J. Zapata (1986) Anther culture in rice. IV. The effect of abscisic acid on plant regeneration. Plant Cell Rep. 5: 136–139.

Tsai, S.C. & M.H. Lin (1977) Production of rice plantlets by anther culture. J. Agric. Res. China 26: 100–112.

Tsay, H.S. & L.J. Chen (1984) The effects of cold shock and liquid medium on callus formation in rice anther culture. J. Agric. Res. China. 33: 24–29.

Tsay, H.S., L.J. Chen, T.H. Tseng & P.C. Lai (1982) The culture of rice anthers of Japoncia × Indica crosses. In: A. Fujiwara (Ed.), Plant Tissue Culture, pp. 561–562. Maruzen Co., Ltd., Tokyo.

Tsay, S.S., J.J. Chen, C.C. Yeh & J.Y. Hsu (1988) Influence of cold shock treatment and microspore developmental stage on rice anther culture. J. Agric. Res. China 37: 257–265.

Tsukahara, M. & T. Hirosawa (1992) Simple dehydration treatment promotes plantlet regeneration for rice (*Oryza sativa* L.) callus. Plant Cell Rep. 11: 550–553.

Uchimiya, H., T. Kameya & N. Takanoshi (1971) *In vitro* culture of unfertilized ovules in *Solanum melongena* and *Zea mays*. Jpn. J. Breed. 21: 247–250.

Vergne, P., F. Riccardi, M. Beckert & C. Dumas (1993) Identification of a 32 kDa marker protein for androgenic response in maize. *Zea mays* L. Theor. Appl. Genet. 86: 843–850.

Wang, C.C. & B.J. Kuang (1981) Induction of haploid plants from the female gametophyte of *Hordeum vulgare* L. Acta Bot. Sin. 23: 329–330.

Wang, C.C., C.S. Sun & Z. Chu (1974) On the condition for the induction of rice pollen plantlets and certain factors affecting the frequency of induction. Acta Bot. Sin. 16: 43–53.

Wang, R.F., Q.X. Zuo, S.W. Zheng & W.Z. Tian (1979) Induction of plantlets from isolated pollen culture in rice *Oryza sativa* subsp. Keng. Acta Genet. Sin. 6: 7.

Wang, Z.X. & N. Iwata (1991) Production of $n + 1$ plants and tetrasomics by means of anther culture of trisomic plants in rice (*Oryza sativa* L.). Theor. Appl. Genet. 83: 12–16.

Wang, Z.Y. & S.D. Tanksley (1989) Restriction fragment length polymorphism in *Oryza* sativa. Genome 32: 1113–1118.

Woo, S.C. & C.C. Chen (1982) Rice anther culture in Taiwan. In: Rice Tissue Cult Planning Conf. IRRI Los Banos, pp. 83–90.

Woo, S.C., S.W. Ko & C.K. Wong (1983) Anther culture of pollen plants derived from cross *Oryza sativa* L. × *O. glaberrima* Steud. Bot. Bull Acad. Sin. 24: 53–58.

Yamaguchi, M., A. Yomoda & K. Hinata (1990) Varietal differences in the response to low temperature treatment for callus formation in anther culture of rice. Jpn. J. Breed. 40: 193–198.

Yang, H.Y. & C. Zhou (1979) Experimental researches on the two pathways of pollen development in *Oryza sativa* L. Acta Bot. Sin. 21: 345–351.

Yang, H.Y. & C. Zhou (1982) *In vitro* induction of haploid plants from unpollinated ovaries and ovules. Theor. Appl. Genet. 63: 97–104.

Yang, X.R., J.R. Wang, H.L. Li & Y.F. Li (1980) Studies on the general medium for anther culture of cereals and increasing of frequency of green pollen plantlets – induction of *Oryza sativa* subsp. Snien. Acta Phytophysiol. Sin. 6: 67–74.

Zapata, F.J. & L.B. Torrizo (1986) Heat treatment to increase callus induction efficiency in anther culture of IR 42. Intl. Rice Res. Newsl. 11: 25–26.

Zapata, F.J., L.B. Torrizo, R.O. Romero & M.S. Alejar (1982) Androgenesis in *Oryza sativa*. In: A. Fujiwara (Ed.), Plant Tissue Culture 1982, Proc. 5th Intl. Cong. Plant Tissue and Cell Culture, Tokyo, July 11–16, pp. 531–532. Maruzen Co., Ltd., Tokyo.

Zapata, F.J, L.B. Torrizo & R.R. Aldemita (1985) Effect of a conditioned medium on callus production and plant regeneration in rice anther culture. Intl. Rice Res. Newsl. 10: 14.

Zapata, F.J., R.R. Aldemita, A.U. Novero, L.B. Torrizo, L.B. Magaling, A.M. Mazaredo, R.M. Visperal, M.S. Lim & H.P. Moon (1986a) IRRI, Korea Collaborative project for the development of cold-tolerant lines through anther culture. IRRI Res. Pap. Ser. 118, 79.

Zapata, F.J., L.B. Torrizo, R.R. Aldemita, A.U. Novero, A.M. Mazaredo, R.M. Visperas, M.S. Lim, H.P. Moon & M.H. Heu (1986b) Rice anther culture. A tool for production of cold tolerant lines. In: Rice Genetics: Proc. Intl. Rice Genetics Symp., 27–31 May 1985, pp. 773–780.

Zapata, F.J., R.R. Aldemita, E.S. Ella & M.S. Cho (1990) Isolated microspore culture of rice at the International Rice Research Institute. In: Rice Genetics II. Proc. Second Intl. Rice Res. Genet. Symp. 14–18 May 1990, pp. 311–319. IRRI, Manila.

Zapata, F.J., M.S. Alejar, L.B. Torrizo, A.U. Novero, V.P. Singh & D. Senadhira (1991) Field performance of anther-culture-derived lines from $F_1$ crosses of Indica rices under saline and nonsaline conditions. Theor. Appl. Genet. 83: 6–11.

Zhang, Z. (1982) Application of anther culture techniques to rice breeding. In: Rice Tissue Culture planning Conf., pp. 55–61. IRRI, Los Banos.

Zhang, Z. (1992) Anther culture for rice breeding at SAAS. In: K. Zheng and T. Murashige (Eds.) Anther culture for Rice Breeders. pp. 38–74. Hangzhou, China.

Zhang, Z.H. & Q.R. Chu (1986) Advances in rice anther culture for varietal improvement in China. J. Agric. Sci. China 2 Suppl.: 10–16.

Zhao, C.Z. & X.F. Qi (1991) The effect of a male gametocide on the culture ability of Indica rice. Acta Agron. Sin. 17: 228–232.

Zhou, C. & H.Y. Yang (1981a) Induction of haploid rice plantlets by ovary culture. Plant Sci. Lett. 20: 231–237.

Zhou, C. & H.Y. Yang (1981b) Embryogenesis in unfertilized embryo sacs of rice. Acta Bot. Sin. 23: 176–180.

Zhou, C. & H.Y. Yang (1981c) Studies on the *in vitro* induction of callus from rice embryo sacs. Hereditas China 3: 10–12.

Zhou, C., H.Y. Yang, H. Yang & S. Cai (1983) Factors affecting callus formation in unpollinated ovary culture of rice. pp. 81–94. In: Proc. of workshop on 'Cell and Tissue Culture Techniques for Cereal Crop Improvement', Science Press Beijing and Int. Rice Research Institute, Manila, Philippines.

Zuo, Q.X., Y. Chen, S.Y. Li & M.G. Wang (1983) Application of gradient centrifugation in the culture of isolated pollens of rice. In: Annual Rep. of the Inst. of Genetics. Academia Sinica 1982, p. 31. Science Press, Beijing.

# 2. *In vitro* haploid production in maize

BERND BÜTER

**Contents**



## 1. Introduction

In 1993, maize (*Zea mays* L.) was produced on 127 million hectares with a worldwide production of about 500,000 metric tons, thus representing the third most important crop after wheat and rice (FAO, 1994). Maize is grown in a wide range of climates and represents a major crop both in industrialized and developing regions of the world. It serves as food for direct human consumption as well as for animal feed. Although rich in metabolizable energy, the nutritional value of maize is limited by inferior protein quality, i.e., a deficiency in lysine and tryptophan.

Maize production is threatened by various pests and pathogens, e.g., European corn borer, maize stalk borer, southern and northern leaf blight, rusts, common smut, downy mildews, etc., as well as by abiotic stresses like cold or drought. Due to their high yield potential, maize hybrids were introduced early in this century. Since then hybrid cultivars have gradually

*S.M. Jain, S.K. Sopory & R.E. Veilleux (eds.),* In Vitro *Haploid Production in Higher Plants, Vol. 4,* 37–71.

advanced to dominate maize production. Breeding maize hybrids normally involves the development of inbred lines by repeated self-pollinations. Starting from heterozygous crosses, 6–8 generations of self-pollinating are required to obtain sufficient homozygosity. During this period first selections on agronomic characters and combining ability may be carried out (i.e., early generation testing). Finally field testing of hybrid crosses accounts for an additional 3–4 generations. Thus, approximately 6–10 years are needed to develop new maize hybrids (depending on the number of growing seasons available per year).

The *in vitro* production of doubled haploid (DH) plants allows for the rapid attainment of homozygous lines, thereby considerably reducing the time required for hybrid development. An acceleration of cultivar release by 2–3 years appears to be a realistic estimation. Since DH plants do not segregate, their use also avoids the risks related to testing in early generations during self-pollination.

Future maize production faces numerous problems and new goals. Therefore, maize breeders must remain flexible and willing to realize new breeding goals in a minimum of time. Along with other *in vitro* techniques, the production of DH plants could become an important tool to support maize breeding. Due to their unique genetic characteristics, DH lines in maize are assumed to be useful not only for hybrid breeding, but also for future applications, e.g., *in vitro* selection for disease and abiotic stress resistance, gene mapping and genetic transformation.

In maize, various approaches have been used successfully to obtain DH plants:

i) Due to abnormal fertilization, maternal haploids may be obtained if the strain Stock 6 (Coe, 1959) or the line WS 14 (Lashermes & Beckert, 1988) are used as male parents (*in situ gynogenesis*).

ii) Analogously, paternal haploids are occasionally induced if the line W 23 (which carries the *ig* = *indeterminate gametophyte* mutation; Kermicle, 1969) is used as the female parent (*in situ androgenesis*).

Using *in vitro* culture methods DH plants have been produced either by

iii) ovary/ovule culture (*in vitro gynogenesis*); or

iv) anther/microspore culture (*in vitro androgenesis*).

*In vitro* gynogenesis so far has been reported only by two groups (Ao *et al.*, 1982; Truong-Andre & Demarly, 1984). More research will be required to evaluate the potential of this method for maize. This section will focus on DH plant production by *in vitro* culture of anthers or isolated microspores.

## 2. Chronology and pathways of androgenic microspore development

In cultures of anthers or microspores, only a relatively small fraction of the microspore population deviates from the normal, gametophytic development. Typically, during the first hours of microspore culture, most of the microspore population becomes swollen and stains brightly with fluorescein diacetate (FDA). Later, i.e., during the first days of culture, rapid abortion of the majority of microspores occurs. Within the first week, the frequency of viable microspores declines from 80–90% to 10% or less (Pescitelli *et al.*, 1990a). Within the viable fraction, some microspores start to undergo mitosis ("induced" microspores = IM). Frequencies of IM reach maximal values after 7–10 days (Pescitelli & Petolino, 1988). After continued mitotic divisions, the induced microspores eventually develop into multinuclear structures which remain within the exine. These structures have been called "putative embryogenic microspores" (PEM; Pace *et al.*, 1987) or early multinuclear structures (early MNS; Gaillard *et al.*, 1992). Visualization of these structures can be performed by specific DNA stains, e.g., DAPI, mithramycin or Feulgen reaction. Later, the enlarged embryogenic structures (intermediate MNS) burst the exine. During this stage cell wall formation starts with short internal wall fragments linked to the intine and continues by centripetal ingrowth of these fragments. The embryogenic structures (late MNS) eventually become completely partitioned and are surrounded by a peripheral cell layer (Gaillard *et al.*, 1992). Finally, multicellular, macroscopically visible embryo-like structures (ES) are obtained.

Induction frequencies in anther and microspore cultures appear to be similar, but are generally lower than in many other species, e.g., barley. Comparisons of IM and ES frequencies indicated that in maize the IM frequency in a given culture may exceed the ES frequency by a factor of 5–10 (Pace *et al.*, 1987). Therefore, methods that reduce IM-abortion very likely will contribute to an improved ES production. Although a single microspore will generally produce a single ES, the formation of multiple ES (2–4) from individual microspores also has been reported (Wan & Widholm, 1992).

Androgenic microspore development in maize may follow different pathways (Miao *et al.*, 1978; Kuo *et al.*, 1986; Pescitelli & Petolino, 1988; Barnabas *et al.*, 1987; Coumans *et al.*, 1989; Gaillard *et al.*, 1991; Pretova *et al.*, 1993; Pescitelli *et al.*, 1994). They can be summarized as follows:

A. No (unequal) first microspore division and no subsequent differentiation of a vegetative and a generative nucleus occur. Instead, the first division gives rise to two similar nuclei which both continue to divide and equally contribute to the developing multinuclear structure.

B. After an unequal first microspore division a large vegetative and a small generative nucleus can be distinguished.

i) The vegetative nucleus starts to divide symmetrically and to develop into a multinuclear embryogenic mass. In anther culture this pathway has been reported to be the most prominent (Barnabas *et al.*, 1987; Pescitelli & Petolino, 1988). Asymmetrical divisions of the vegetative nucleus also have been observed which lead to the formation of multinuclear masses without subsequent cell wall formation. Since no further development was observed in these multinuclear structures, they were assumed to be developmental "dead ends" (Pescitelli *et al.*, 1994).

ii) The embryogenic microspore originates from the generative nucleus. Multicellular embryogenic structures produced during this pathway typically contain smaller cells and nuclei that stain more densely than in structures originating from the vegetative nucleus. In cultures of isolated microspores this pathway was found to be dominant (Coumans *et al.*, 1989; Pretova *et al.*, 1993; Pescitelli *et al.*, 1994).

iii) The vegetative and generative nucleus fuse and give rise to a large microspore with one enlarged nucleus and no vacuole. According to Pescitelli and Petolino (1988), about 15% of the induced microspores in maize anther culture may belong to this category. However, with the exception of rarely occurring nuclear divisions, no further development into multinuclear or multicellular structures occurred.

iv) Infrequently, both the vegetative and the generative nucleus start to divide and to develop independently from each other, thus forming a multicellular embryogenic structure which consists of two distinct types of cells, i.e., large cells with diffuse staining nuclei and small cells with densely staining nuclei (Pretova *et al.*, 1993).

## 3. Anther culture

### 3.1. *Procedures*

Immature tassels are removed from the donor plants before emergence from the leaf whorl. Usually a cold treatment is applied to the tassels before culture initiation. After the cold treatment anthers containing microspores in the mid- to late-uninucleate or early-binucleate stage are excised from the florets and inoculated on an induction medium. Anthers with microspores in theses stages are supposed to be most suitable for *in vitro* androgenesis. After approximately 21–28 days the first microspore-derived, embryo-like structures (ES) will be observed (Fig. 1). The ES are subsequently transferred to regeneration medium which supports the development of plantlets (direct regeneration). These plantlets are grown to maturity and, provided chromosome doubling has occurred, can be self-pollinated to obtain doubled haploid seed. In the following discussion, the capacity of a specific donor plant or genotype to form ES from which plants can be regenerated is referred to as androgenic potential or androgenic response.

In an alternative approach, ES can be used for the induction of regenerable, totipotent callus (Pescitelli *et al.*, 1989). In order to establish regenerable callus ES are transferred to media containing either 2,4-D or dicamba (2.0–2.5 mg/L) and L-proline in relatively high concentrations (2.5–3.0 g/L). Upon transfer to regeneration medium these regenerable calli will develop

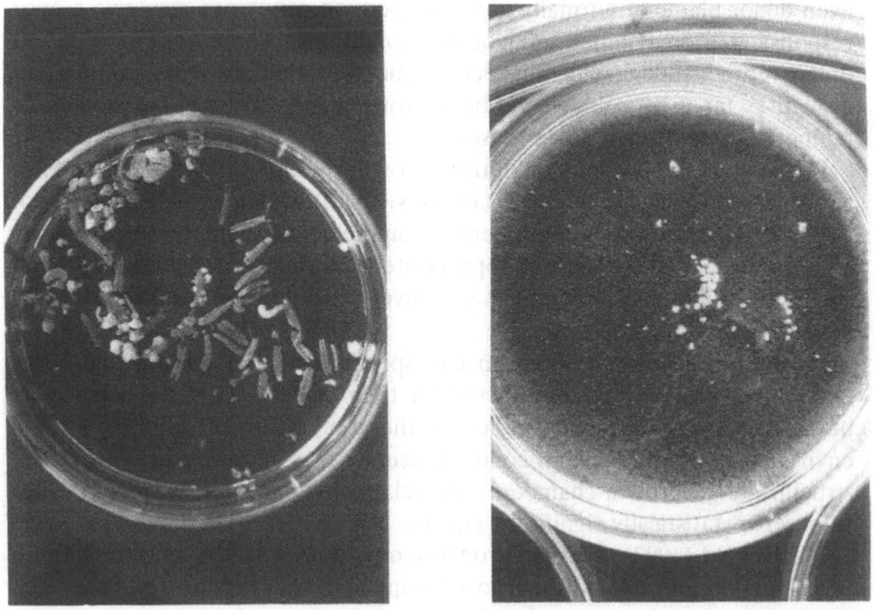

*Figure* 1. Formation of embryo-like structures (ES) in maize anther (left) and microspore culture (right).

into plantlets (indirect regeneration) similar to plantlets which have been regenerated directly. Such indirect regeneration may facilitate the production of several plants from one regenerable callus, thus allowing sib-pollination if selfing cannot be accomplished. Another advantage of the intermediate callus phase is its suitability for the induction of chromosome doubling (see 3.2.7). However, in order to avoid chimeric regenerants, induction of chromosome doubling appears to be more practical during anther culture before nuclear or cell divisions occur. Furthermore an extended *in vitro* culture duration that results from an additional callus phase might increase the risk of induced genetic variation as reported for other species (Ziauddin & Kasha, 1990).

## 3.2. *Factors affecting anther culture*

### 3.2.1. *Conditions during donor plant growth*

Differences in culture efficiency between field- and greenhouse-grown plants or between plants grown in different seasons are common in maize anther culture (Dieu & Beckert, 1986; Genovesi, 1990; Barloy & Beckert, 1993). It indicates that the quality of the donor plants is critical with regard to subsequent anther culture response. Personal experience has shown that even donor plants of normally highly responsive genotypes may give a poor response if grown under unfavorable conditions. Therefore, in order to obtain reproducible results, it is recommended to standardize the donor plant growth as far as possible, e.g., by growing plants in growth chambers or temperature-controlled greenhouses.

Principally it is believed that optimal growth conditions will support *in vitro* androgenesis (Nitsch *et al.*, 1982; Genovesi, 1990). Day/night temperatures of 25–30/18–20°C, 16–18 h photoperiod, supplemental light preferably from high pressure sodium lights, appropriate watering and fertilization, and avoidance of pesticide treatments whenever possible have been found to be advantageous.

In spite of the alleged importance of specific growth conditions, no systematic studies have been conducted on their effects. This deficiency may be related to the fact that studies on the comparison of different growth environments are difficult to conduct: usually they require expensive infrastructure, e.g., growth chambers, and relatively large sample sizes in order to observe statistically significant effects.

MacDonald (1992) recorded growth temperatures of donors throughout a 2-year period and related them to subsequent anther culture response. The results led her to the assumption that the minimal growth temperature occurring during tassel development had a specific impact on ES production. She therefore recommended that temperature should not fall below 17°C during two weeks prior to tassel harvest. However, since temperature probably was only one factor among several (light quality, photoperiod) that varied, it was difficult to draw unequivocal conclusions on the temperature effect. On the other hand, for culture of isolated microspores using different donor plant genotypes, Afele *et al.* (1992) reported beneficial effects of reduced temperatures (18/15°C) before tassel harvest. Clearly, there exists a need for further studies including a broader range of genotypes grown under controlled environments as well as potential interactions (e.g., genotype × growth environment or growth environment × pre-culture treatments) in order to understand this complex issue.

In other crops, e.g., wheat, treating donor plants with gametocidal agents was found to improve anther culture response significantly (Schmid & Keller, 1986). Similarly in maize, the application of a specific gametocide ("CGA") was found to increase ES production (Büter *et al.*, 1990); optimal effects were obtained when the gametocide was applied twice, i.e., immediately before and after the meiosis in pollen mother cells (Büter, 1992).

3.2.2. *Pre-inoculation temperature stress*

As reported for other crops (Henry & De Buyser, 1990; Flehinghaus *et al.*, 1991; Hou *et al.*, 1993) treatment of anthers/tassels at reduced temperature before culture initiation seems to be essential for efficient *in vitro* androgenesis in maize. Cold stress is thought to disrupt normal, gametophytic development and to stimulate the sporophytic pathway.

Although the exact mode of action still remains unclear, evidence has been presented that, at least in some genotypes, a specific protein may be involved in the cold temperature effects. This protein (MAR32) with a molecular mass of 32 kDa is induced and accumulates in maize anthers during cold pretreatment (Vergne *et al.*, 1993). In individual tassels a positive correlation was observed between the amount of MAR32 after a 7-day cold treatment and subsequent ES production. MAR32 thus represented a marker for androgenic capacity in these tassels. However, although one genotype included in the study was responsive in anther culture, it did not accumulate MAR32. This indicates that MAR32 probably is not a generally applicable marker, but is restricted to a certain form of androgenic responsiveness in maize. Furthermore no MAR32 accumulation was detected within microspores isolated from MAR32-containing anthers; therefore it was concluded that the protein is involved in a sporophytically determined competence for androgenesis.

Cold-pretreated microspores during the first 30 days of culture were found to have a lower total polyamine content compared to gametophytically developing, *in situ* microspores (Santos *et al.*, 1995). Although only one genotype was investigated in this study, it was suggested that exogenous application of polyamines or polyamine-inhibitors to the induction medium might facilitate an improved androgenic response, especially in recalcitrant genotypes.

Various pretreatment durations and temperatures have been recommended for optimal androgenic response, indicating that the cold pretreatment might interact with other factors, e.g., genotype, temperature during donor plant growth or *in vitro* culture temperature. Typically cold pretreatments in maize are applied for 7–14 days at temperatures ranging from 4 to 14°C (Nitsch *et al.*, 1982; Genovesi & Collins, 1982; Dieu & Beckert, 1986; Tsay *et al.*, 1986).

3.2.3. *Post-inoculation temperature stress*

In *Brassica napus*, culture temperature during the initial hours after inoculation is critical for the induction of the androgenic pathway. It was shown that temperature can be used as a simple experimental variable to obtain either sporophytic (32.5°C) or gametophytic (17.5°C) microspore development (Custers *et al.*, 1994).

Unfortunately, in maize, the effects of post-inoculation temperature stress appear to be less dramatic. Nevertheless, considerable increases in ES and plant production have been accomplished both by cold (Nitsch *et al.*, 1982; Dieu and Beckert, 1986; Deimling *et al.*, 1990) and heat treatments (Genovesi, 1990). Whether reduced (e.g., 10–14°C) or elevated (e.g., 30–32°C) temperatures are more suitable may depend on various cultural parameters, e.g., genotype, pretreatment, donor plant growth, and therefore individual analysis for each genotype and experimental protocol is required (Büter *et al.*, 1991). In our laboratory, good results in a broad range of genotypes have been obtained after a 7-day post-plating treatment at 14°C. The bene-

*Table* 1. Basic media formulations most frequently used in maize anther culture

| Components [mg/L] | N6 (CHU, 1981) | YÜ-PEI (Ku *et al.*, 1981) |
|---|---|---|
| $NH_4NO_3$ | – | 165 |
| $(NH_4)_2SO_4$ | 463 | – |
| $KNO_3$ | 2830 | 2500 |
| $KH_2PO_4$ | 400 | 510 |
| $MgSO_4 \cdot 7H_2O$ | 185 | 370 |
| $CaCl_2 \cdot 2H_2O$ | 166 | 176 |
| $MnSO_4 \cdot 4H_2O$ | 4.4 | 4.4 |
| $ZnSO_4 \cdot 7H_2O$ | 1.5 | 1.5 |
| $H_3BO_3$ | 1.6 | 1.6 |
| KI | 0.8 | 0.8 |

ficial effects of short term cold treatments were found to be further increased if the treatments were combined with the addition of L-proline to the induction medium (Büter *et al.* 1991).

### 3.2.4. *Induction media*

Culture media used for the induction of sporophytic microspore development and subsequent production of embryo-like structures usually are referred to as induction media. In the early years of maize anther culture, Chinese colleagues spent much effort on the evaluation of various *basic media formulations* (Chu, 1981; Ku *et al.*, 1981). Following these studies media based on either N6 (Chu, 1981) or YÜ-PEI (Ku *et al.*, 1981) formulations (Table 1) supplemented with FeEDTA ($10^{-4}$ M) most commonly have been used for maize (Brettel *et al.*, 1981; Genovesi & Collins, 1982; Pauk, 1985; Dieu & Beckert, 1986; Petolino & Jones, 1986; Tsay *et al.*, 1986; Pace *et al.*, 1987).

With regard to the *carbohydrate supply*, sucrose seems to provide the best results in maize. Concentrations ranging from 6 to 15% have been recommended. Our own experiments concerning sucrose concentration revealed 9% to be optimal (Table 2); however for some genotypes, 12% appeared best (Genovesi, 1990).

In some species, e.g., barley or wheat, maltose and occasionally glucose or cellobiose were found to be superior to sucrose in terms of embryo and plant production (Hunter, 1987; Finnie *et al.*, 1989; Chu *et al.*, 1990; Orshinsky *et al.*, 1990; Cai *et al.*, 1992). However, in maize anther culture, similar results were not obtained (Büter *et al.*, 1995). A comparison of eight carbohydrates as components of the induction medium revealed that sucrose resulted in significantly higher overall plant production than all other sugars tested. Only raffinose gave comparable ES production, while plant regeneration was considerably lower (Table 3). Superiority of sucrose compared to maltose or other carbohydrates has also been suggested by other authors (Genovesi, 1990; Pescitelli *et al.*, 1994). In contrast, some sweetcorn cultivars

*Table* 2. Comparison of different sucrose concentrations in induction medium for maize anther culture[1] (Büter, unpublished)

| Sucrose conc. [%] | ES[2]/100 anthers | RP[3]/100 anthers |
|---|---|---|
| 6 | 55.6 b[4] | 1.4 b[4] |
| 9 | 107.7 a | 7.7 a |
| 12 | 77.1 b | 5.2 ab |
| 15 | 66.7 b | 4.1 ab |

[1] Genotype: ETH-M 24 (= DH [DAN SAN 91] × DH [PA91 × FR16]; 480 anthers/treatment; sucrose autoclaved.
[2] ES = embryo-like structures.
[3] RP = regenerated plants.
[4] Treatments followed by the same letter are not significantly different from each other (LSD: 0.05).

*Table* 3. Comparison of various carbohydrates for their aptitude in maize anther culture[1] (Büter *et al.*, 1995)

| Carbohydrates | ES[2]/100 anthers | RP[3]/100 anthers |
|---|---|---|
| Sucrose | 122.0 a[4] | 8.9 a[4] |
| Raffinose | 118.2 a | 4.2 b |
| Cellobiose | 54.5 b | 3.5 bc |
| Maltose | 37.8 bc | 0.9 bc |
| Fructose | 20.8 bc | 2.7 bc |
| Glucose | 7.4 c | 1.5 bc |
| Trehalose | 2.1 c | 0 c |
| Galactose | 1.2 c | 0 c |

[1] Genotype: ETH-M 24 (= DH [DAN SAN 91] × DH [PA91 × FR16]; 336 anthers/treatment. Carbohydrates were applied in liquid, activated charcoal-containing induction medium at a concentration of 0.26 M. Media were sterilized by autoclaving.
[2] ES = embryo-like structures.
[3] RP = regenerated plants.
[4] Treatments followed by the same letter are not significantly different from each other (LSD: 0.05).

were found to give better response on maltose-containing (13%) induction medium (MacDonald, 1992).

Not only the choice of carbohydrate, but also the *method of media sterilization*, i.e., autoclave or filter-sterilization, was found to have an effect on anther culture. Filter-sterilization was reported to provide increased ES frequencies compared to autoclaving the induction medium (MacDonald, 1992; Büter *et al.*, 1993). This was probably due to the formation of inhibitory substances during hydrolysis of sucrose (or other carbohydrates) when subjected to high temperatures that occur during autoclaving (Weatherhead *et al.*, 1979; Theander & Nelson, 1988). Consequently, autoclaving FeEDTA, a catalyst of sucrose breakdown, separately from the remaining medium

*Table* 4. Carbohydrates as components of the induction medium: impact on maize anther culture efficiency[1] (Büter *et al.*, 1995)

| Carbohydrates [g/L] | ES[2]/100 anthers | RP[3]/100 anthers |
|---|---|---|
| Su + Glu + Fru [27 + 33 + 33] | 125.0 a[4] | 6.4 a[4] |
| Su + Glu + Fru [45 + 23 + 23] | 117.6 a | 6.8 a |
| Su + Glu + Fru [63 + 14 + 14] | 105.4 a | 3.2 b |
| Su [90] | 68.6 b | 4.4 b |
| Glu [95] | 27.1 c | 1.5 c |
| Fru [95] | 20.5 c | 1.5 c |

[1] Genotype: ETH-M 24 (= DH [DAN SAN 91] × DH [PA91 × FR16]; 360 anthers/treatment. Carbohydrates (media) in treatments with three sugars were filter-sterilized, carbohydrates (media) in treatments with a single sugar were autoclaved in the presence of activated charcoal.
[2] ES = embryo-like structures.
[3] RP = regenerated plants.
[4] Treatments followed by the same letter are not significantly different from each other (LSD: 0.05).

components resulted in a considerable increase in ES production (Büter *et al.*, 1993).

However, even better effects have been obtained when *activated charcoal* (AC) was added to the induction medium prior to autoclaving. By the addition of 0.5% AC to a liquid induction medium (followed by its removal 24 h after autoclaving) ES production increased from 42.7 (autoclaved without AC) and 52.6 (filter-sterilized) to 111.5 ES/100 anthers (Büter *et al.*, 1993). Similar effects of AC were observed when sucrose was replaced by maltose or glucose (Büter, unpublished). Therefore the primary benefit of AC appears to be the adsorption of inhibitory substances in the culture medium which are produced during the autoclaving process.

The question arises as to why media autoclaved in the presence of AC resulted in much higher ES frequencies than filter-sterilized media. It was assumed that the additional presence of glucose and fructose in the induction medium, as occuring after partial sucrose hydrolysis, would contribute to the better performance of anthers in the autoclaved medium. In fact, it has been shown that filter-sterilized induction media containing a mixture of sucrose, glucose and fructose led to significantly higher frequencies of ES and plant production than media containing either sucrose, glucose or fructose alone (Table 4).

The effects of *plant growth regulators* in maize anther culture have been controversial. Various studies indicate that certain combinations of auxins (particularly 2,4-D: 1–2 mg/L) and cytokinins (particularly kinetin or BA: 0.5–2 mg/L) in the induction medium promote sporophytic microspore development (Miao *et al.*, 1978; Ting *et al.*, 1981; Kuo *et al.*, 1986). Recently, in an extensive study comparing nine different combinations of growth regulators, the highest induction frequencies were found on YÜ-PEI medium supplemented with 2 mg/L 2,4-D and 1.5 mg/L kinetin (Hongchang *et al.*,

*Table 5.* Maize anther culture: effects of amino acids and casein hydrolysate on culture efficiency[1] (Büter, unpublished)

| Amino acids [mg/L] | ES[2]/100 anthers | RP[3]/100 anthers |
|---|---|---|
| Glu + Asp [125 + 15] | 137.1 a[4] | 16.3 a[4] |
| Casein hydrol. [500] | 91.2 bc | 12.4 ab |
| Control (no amino acids) | 70.3 c | 7.9 b |

[1] Genotype: ETH-M 24 ( = DH [DAN SAN 91] × DH [PA91 × FR16]; 440 anthers/treatment. Amino acids and casein hydrolysate were filter-sterilized and then added to the (autoclaved) induction media.
[2] ES = embryo-like structures.
[3] RP = regenerated plants.
[4] Treatments followed by the same letter are not significantly different from each other (LSD: 0.05).

1991). Unfortunately no control treatment without growth regulators was included in this study. It has alternatively been proposed that the level of endogenous growth regulators in maize anthers meets the requirement for ES formation; therefore, no exogenous supply is necessary (Nitsch *et al.* 1982; Tsay *et al.*, 1986). In corresponding experiments supplementation of growth regulators decreased ES production. However, 2,3,5-triiodobenzoic acid (TIBA), a chemical known to act as an antiauxin, was found to give positive results and therefore is widely applied (Ku *et al.*, 1981; Nitsch *et al.*, 1982; Dieu & Beckert, 1986).

Intensive studies have been carried out concerning the impact of specific *organic additives*, e.g., casein, lactalbumin, amino acids, etc., on maize anther culture (Ku *et al.*, 1981; Miao *et al.*, 1978; Nitsch *et al.*, 1982; Genovesi & Collins, 1982; Genovesi, 1990). Among these, casein hydrolysate (CH) was found to improve ES production. Although potentially useful for practical application, CH, as a complex, not clearly defined compound, may not be appropriate for some research purposes, since reproducibility cannot be ensured. Our own studies indicated that the beneficial effect of CH may be related to L-glutamine and L-asparagine. Induction media supplemented by these two amino acids (L-glutamine: 125 mg/L; L-asparagine: 15 mg/L) produced higher numbers of ES than similar media with CH (500 mg/L) or without amino acids (Table 5).

Agar (0.7–0.8%) frequently has been used as a *gelling agent* in induction media (Genovesi & Collins, 1982; Tsay *et al.*, 1986; Petolino & Thompson, 1987). Alternatively, agarose (0.6–0.7%) also has been applied with good success (Nitsch *et al.*, 1982; Dieu & Beckert, 1986). Our own studies on the suitability of different gelling agents revealed best results for media solidified with phytagel (Gelrite) (0.1–0.2%); in comparison to media solidified with agar, gelrite-solidified media resulted in a twofold increase in ES-production. Highest ES frequencies have been observed on liquid media in our laboratory and elsewhere (Pescitelli *et al.*, 1989).

3.2.5. *Plant regeneration*

Regeneration of plants can be accomplished either directly by transferring ES to germination (regeneration) media (Fig. 2) or indirectly by inducing totipotent calli and subsequently transferring them to regeneration media (see 3.1). Direct regeneration has been most frequently applied in maize. Only scant attention has been paid to the optimization of media used for induction of totipotent callus. Pescitelli *et al.* (1989) used a *callus induction medium* consisting of N6 major salts and vitamins, B5 minor salts, 30 g/L sucrose, 2.9 g/L L-proline, 100 mg/L casein hydrolysate, 100 mg/L myo-inositol, 2.5 mg/L dicamba, 0.1 mg/L 2,4-D and 8.0 g/L agar. Increased frequencies of totipotent calli were observed when agar was replaced by phytagel (2.7 g/L) (Fig. 3).

*Regeneration media* for maize are usually N6 or YÜ-PEI based (see 3.2.4), although MS based media have also been used (Ku *et al.*, 1981). Regeneration media differ from induction media mainly by reduced sucrose concentration (2–4%) and by the supplementation with cytokinin, e.g., kinetin (1–2.5 mg/L) or BA (0.5–1.0 mg/L) to promote shoot induction. Kuo *et al.* (1986) reported kinetin (1 mg/L) to be most efficient for plant differentiation. Other authors have suggested the use of media free of growth regulators for regeneration (Petolino & Genovesi, 1994). Efficient plant regeneration seems to require solidified media. Comparisons of different regeneration media conducted in our laboratory revealed that efficient plant regeneration, i.e., 10–20 plants/100 ES, can be accomplished on YÜ-PEI based media containing 30 g/L sucrose and 2.5 mg/L kinetin solidified by agarose (Büter *et al.*, 1995).

In contrast to callus induction, plant regeneration requires exposure of ES to light. Typically ES or regenerable calli are cultured under white fluorescent light (16 h photoperiod) at 26–28°C. Although no clear experimental evidence is available, the addition of red fluorescent light has been suggested to support better plant regeneration (Nitsch *et al.*, 1982; Büter, 1992). Water condensation within the culture dishes usually has an adverse effect on plant regeneration and, therefore, should be avoided.

Transfer of plants to *ex vitro* conditions can be done as soon as the regenerated plants have produced a well-developed root system. Usually, rooting occurs spontaneously either on the regeneration medium or after transfer to growth regulator-free culture medium. In contrast to other species (e.g., barley, wheat), regeneration of albino plants occurs infrequently and thus is not considered a major problem in maize.

3.2.6. *Time of transfer to regeneration medium*

In the standard procedure ES 1–2 mm in diameter are transferred to regeneration medium in order to stimulate shoot/plant formation. However, in wheat it was observed that a transfer of anthers to the regeneration medium as early as 14 days after culture initiation, i.e., prior to the macroscopic appearance of ES, does not increase embryogenesis, but does significantly

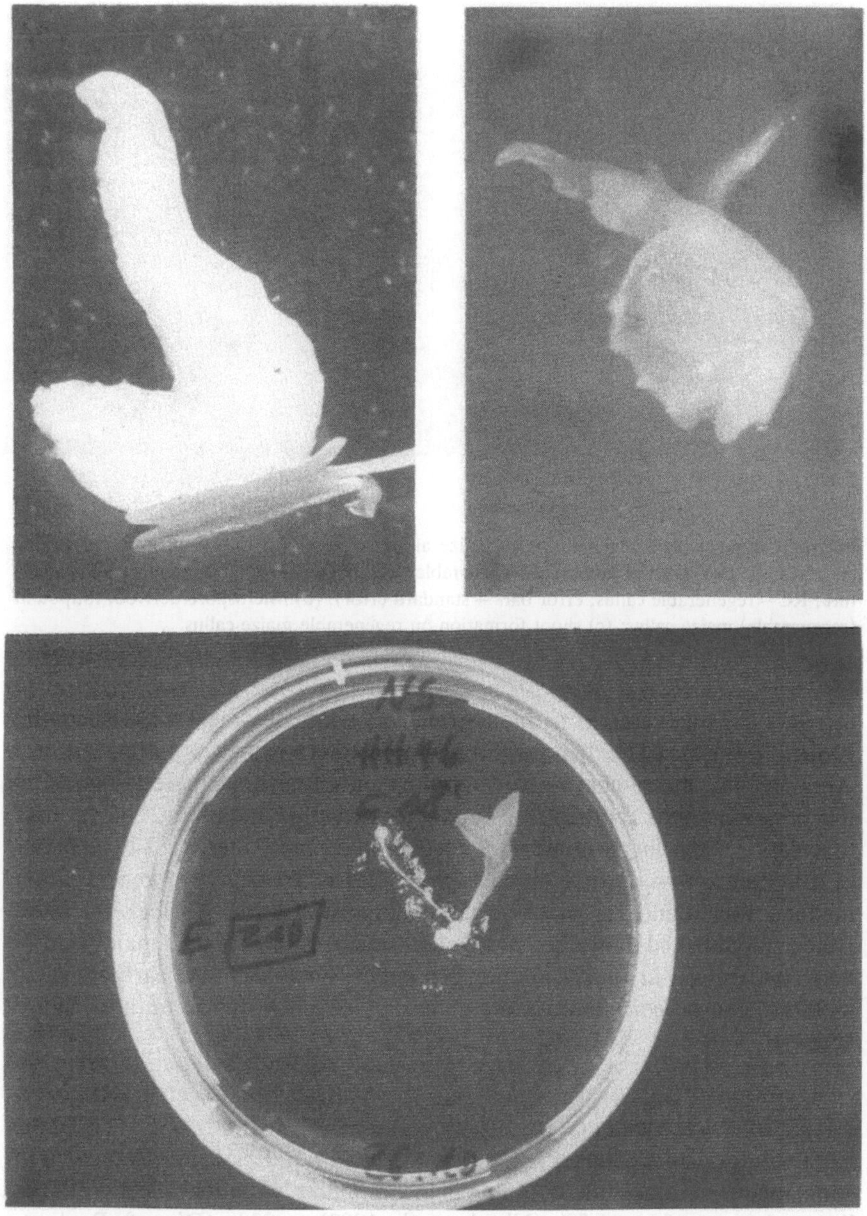

*Figure* 2. Direct plant regeneration in maize anther culture: germinating embryo (above left); shoot and root formation (above right); subculture of regenerated plant (below).

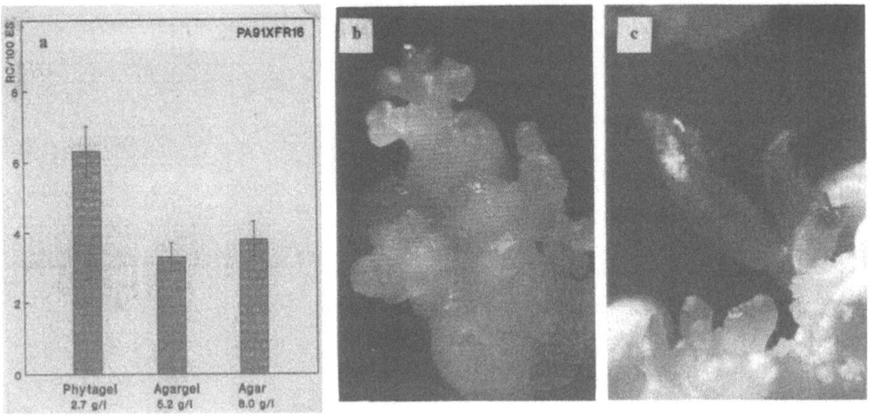


*Figure* 3. Indirect plant regeneration in maize anther culture: (a) Impact of different gelling agents on the formation of totipotent (regenerable) calli (Büter, 1992) (ES = embryo-like structure; RC = regenerable callus; error bars = standard error); (b) Microspore-derived, totipotent (regenerable) maize callus; (c) shoot formation on regenerable maize callus.

increase the regeneration of plants (Henry & De Buyser, 1981). Recently, Barloy & Beckert (1993) reported a similar effect in maize anther culture. Although the mean number of embryogenic anthers decreased slightly in the best treatment (anther transfer to regeneration medium 21 days after inoculation), the mean number of regenerated plants increased by a factor of 10. Later, this positive impact of an "early transfer" to regeneration medium was found for many other genotypes (Barloy & Beckert, 1993; Büter, unpublished). Insufficient regeneration of ES has been a major bottleneck, particularly in low responding genotypes. Therefore the "early transfer-method" can be considered a major breakthrough in maize anther culture (Fig. 4).

### 3.2.7. *Chromosome doubling*

Compared to other crops, e.g., barley (Lyne *et al.*, 1986), the frequency of spontaneous chromosome doubling is low in maize. Frequencies cited in the literature range from 4.5% (Nitsch *et al.*, 1982) to 22% (Dieu & Beckert, 1986), with a majority of studies reporting frequencies around 10%. Differences in the rate of spontaneous chromosome doubling are believed to originate from different genetic competencies of the donor plant genotypes.

Chromosome counts often reveal lower frequencies of DH plants than counts based on the number of fertile plants, i.e., plants which set seed; this indicates that plants reported as fertile in some cases may have been

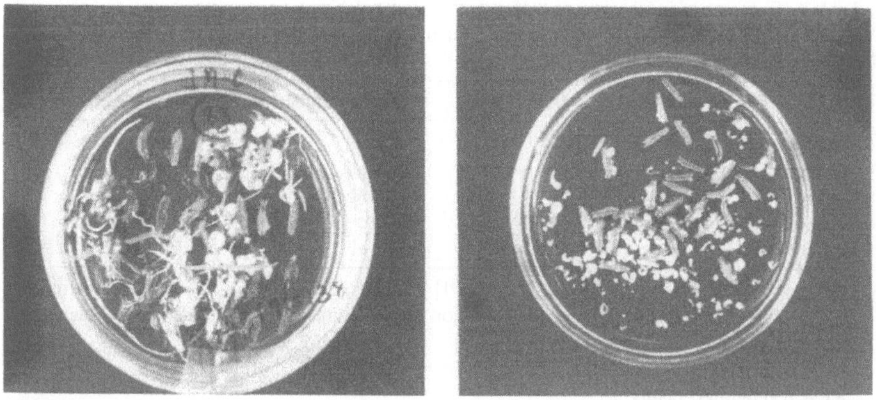

*Figure* 4. Comparison of plant formation in culture dishes with (left) and without (right) early anther transfer.

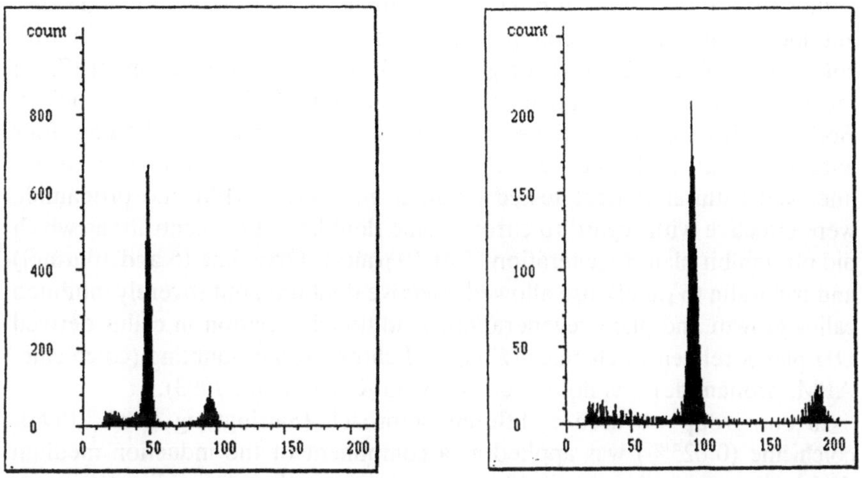


*Figure* 5. Flow cytometer histogram of a haploid (left) and doubled haploid, microspore-derived maize plant (right); nuclei were stained with 4′, 6-diamidino-2-phenylindole (DAPI); x-axis: histogram channels revealing the DNA content in relative fluorescence units; y-axis: number of nuclei counted per histogram channel.

polysomatic chimeras with diploid sectors in tassels and ears. Flow cytometry has been used to confirm the ploidy status of microspore-derived plants reliably (Fig. 5).

Unlike other monocots, maize plants do not normally produce tillers. For this reason, *in situ* doubling treatments applied to regenerated plants have not been successful. Attempts to induce chromosome doubling in maize

*Table* 6. Temporary incubation of maize anthers on colchicine-containing induction medium: impact on the frequencies of embryo-like structures, regenerated plants and doubled haploid plants[1] (Saisingtong *et al.*, 1996)

| Colchicine [duration] | ES[2]/100 anthers | RP[3]/100 anthers | DH[4]/100 anthers |
|---|---|---|---|
| Colchicine [1 day] | 199.0 a[5] | 40.6 a[5] | 2.7 b[5] |
| Colchicine [3 days] | 167.2 ab | 29.6 abc | 7.0 ab |
| Colchicine [5 days] | 138.7 bc | 21.8 bc | 3.2 b |
| Colchicine [7 days] | 75.2 c | 16.7 c | 9.9 a |
| No Colchicine | 167.2 ab | 31.7 ab | 1.5 b |

[1] Genotype: ETH-M 36 (= DH [DAN SAN 91] × DH [PA91 × FR16]; 456 anthers/treatment; 47–58 plants per treatment were analyzed for ploidy status by flow cytometry. Colchicine concentration: 250 mg/L; temperature: 14°C).
[2] ES = embryo-like structures.
[3] RP = regenerated plants.
[4] DH = doubled haploid plants.
[5] Treatments followed by the same letter are not significantly different from each other (LSD: 0.05).

anther culture have concentrated on the application of colchicine and other antimitotic agents to totipotent calli. Wan *et al.* (1989) treated calli with colchicine (0.025 or 0.05%) for 24, 48 or 72 h; in the best treatment (72 h) more than 90% of the plants showed the diploid chromosome number. Both colchicine concentrations gave similar results. In a second study, four antimicrotubule herbicides (aminoprophosmethyl (APM), pronamide, oryzaline, and trifluralin) were tested (Wan *et al.*, 1991). APM and pronamide were effective with regard to chromosome doubling at concentrations which did not inhibit plant regeneration (5 or 10 μmol). Oryzaline (5 and 10 μmol) and trifluralin (5 μmol) also allowed effective doubling, but severely inhibited callus growth and plant regeneration. Additional variation in callus-derived DH plants related to chemically induced chromosome doubling (colchicine, APM, pronamide) was not observed (Wan & Widholm, 1993).

In our group we used a different approach (Saisingtong *et al.*, 1996): colchicine (0.025%) was applied as a component of the induction medium during the post-inoculation cold treatment, i.e., anthers were directly inoculated on colchicine-containing media (Table 6). Durations of the colchicine application were varied from 1 to 9 days. Optimal frequency of DH plants was obtained when anthers were kept on colchicine medium for 7 days before transfer to colchicine-free induction medium. With this method 49% of the regenerated plants were DH plants compared to 3% in a control treatment without colchicine. In summary, the colchicine treatment led to a production of eight DH plants/100 plated anthers whereas, in the control treatment, two DH plants/100 anthers were obtained.

## 4. Culture of mechanically isolated microspores

### 4.1. *Procedures*

First attempts to establish cultures of isolated maize microspores were made by Nitsch (1977). Although multicellular structures were obtained, no plants could be regenerated from these structures. Successful plant regeneration was first reported in 1989 (Coumans *et al.*, 1989; Pescitelli *et al.*, 1989). Donor plant growth, tassel harvest, tassel cold pretreatment and optimal developmental stage for microspore cultures basically resemble those described for anther culture.

Mechanical isolation of microspores has been carried out using various techniques including macerating anthers against a stainless steel sieve with a glass rod (Pescitelli *et al.*, 1989), chopping anthers with a razor blade (Genovesi & Yingling, 1990), blending anthers in an electric blender (Coumans *et al.*, 1989; Pescitelli *et al.*, 1990a; Büter, 1992) or pulverizing anthers with a dispersing tool (Gaillard *et al.*, 1991). Anthers used for isolation either have been precultured or used directly after the cold pretreatment, i.e., without preculture ("ab initio" cultures). Instead of anthers, spikelets or tassel segments have been used as explants (Coumans *et al.*, 1989; Bedinger & Edgerton, 1990; Gaillard *et al*, 1991; Vergne *et al.*, 1991).

Media specific for isolation of microspores (isolation media) are generally used. Additional steps include the separation of microspores from anther wall debris by filtering and/or centrifugation and the removal of deleterious substances released by the surrounding somatic anther tissue or aborted microspores by washing the microspores repeatedly in fresh isolation or culture medium. Separation of aborted from living microspores and collection of specific microspore stages as homogeneous populations by density centrifugation, e.g., using percoll or sucrose gradients (Fig. 6), also were reported to be beneficial (Pescitelli *et al.*, 1989, 1990a; Gaillard *et al.*, 1991).

After isolation, microspores are suspended in induction media similar to those used for anther culture. Embryo-like structures may be observed 14–21 days after culture initiation (Fig. 1), i.e., ES formation in microspore cultures seems to occur somewhat earlier than in anther culture (see 3.1). Regeneration and maturation of plants follow the procedures described for anther culture.

### 4.2. *Factors affecting isolated microspore culture*

#### 4.2.1. *Pre-inoculation temperature stress*

Standard procedures for microspore culture involve a tassel cold treatment at 8–10°C for 14 days (Coumans *et al.*, 1989; Genovesi & Yingling, 1990; Afele *et al.*, 1992). The late-uninucleate to early-binucleate stage of microspore development appear to be optimal (Gaillard *et al.*, 1991; Pescitelli *et*

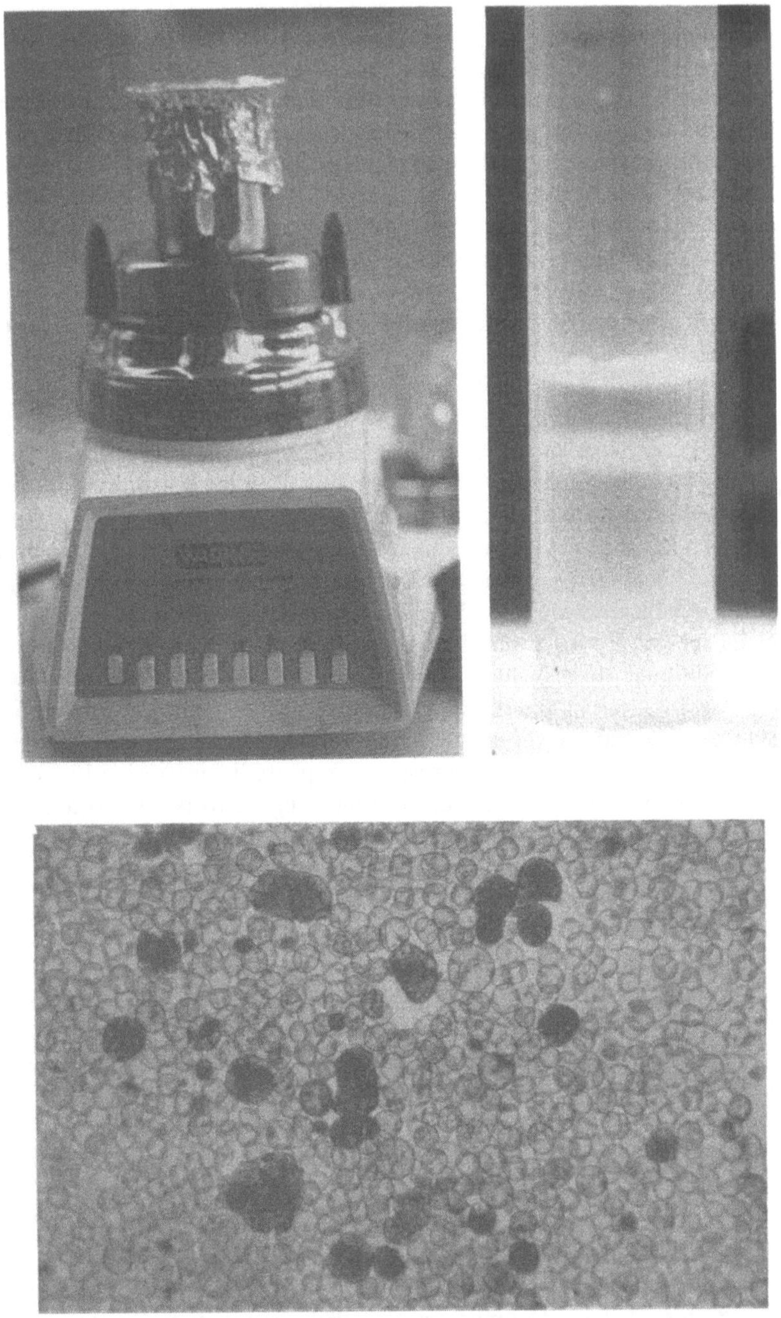

*Figure* 6. Microspore culture: blender isolation (above left), viable microspores collected in a Percoll gradient (above right), microspore suspension during induction phase (below).

*al.*, 1989, 1994). Generally, compared to anther culture there seems to exist a tendency towards more advanced stages (Pescitelli *et al.*, 1994). Gaillard *et al.* (1991) reported a beneficial effect of an extended cold pretreatment of tassels before microspore isolation: Tassel pretreatment at 7°C for 17–18 days resulted in an increased frequency of induced microspores and embryo-like structures compared to 12–13 or 14–16 day treatments. However, the study did not address whether this was caused by a specific cold-related, physiological effect or rather by the advanced developmental stage of the microspores caused by the prolonged cold treatment. In fact, in the same study, late-uninucleate to early-binucleate microspores were found to produce more ES than mid- to late uninucleate microspores.

#### 4.2.2. *Preculture of anthers/spikelets*

Enhanced embryo formation in microspore cultures has been observed after preculture of anthers/spikelets. Pescitelli *et al.* (1989) precultured anthers on induction medium for 3 or 7 days. Microspores isolated from precultured anthers gave a 5- and 16-fold higher ES formation than non-precultured anthers. For improved response in isolated microspore cultures, Genovesi & Yingling (1994) suggested a shortened cold treatment to tassels (4 days at 10°C) followed by a 10–14 day anther preculture at 10°C on a mannitol solution.

#### 4.2.3. *Microspore isolation*

Detailed descriptions of the mechanical isolation of microspores have been published but only a few studies have compared different *isolation techniques*, making it difficult to determine if one technique is more suitable than another. In addition it is possible that different donor materials (genotypes, growth conditions) require different isolation techniques in order to obtain optimal response.

A study comparing microspore isolation by maceration of anthers, i.e., pressing anthers against a stainless steel sieve, and by blender isolation indicated that blending is less stressful for the microspores (Pescitelli *et al.*, 1989). Gaillard *et al.* (1991) compared isolation by blending vs. a dispersing tool (Ultra Turrax). Microspore viability after dispersing was 85–95% compared to less than 75% after blending. However, Büter (1992) reported initial viability frequencies greater than 90% using blender isolation. Different donor plant genotypes or different handling of the isolation equipment (speed, duration) may have caused this discrepancy.

Few explicit data are available concerning the optimal *explant type* (anthers, spikelets, tassel segments) for microspore isolation. In our experiments using blender isolation we have found that treatments using spikelets as explants were more efficient in terms of ES formation than treatments using anthers (Table 7). In addition, using spikelets or tassel segments considerably reduced the time required for isolation, since no anthers needed to be excised

*Table* 7. Culture of isolated maize microspores: Impact of explant type and blending protocol on ES production[1] (Nägeli & Büter, unpublished)

| Explant [Speed of blending (RPM)] | ES[2]/100 anther equivalents[3] |
|---|---|
| Florets [10,900] | 163.9 a[4] |
| Florets [16,700] | 103.2 b |
| Florets [21,000] | 99.5 b |
| Anthers [6,700] | 11.5 d |
| Anthers [10,900] | 69.1 bc |
| Anthers [16,700] | 16.4 cd |

[1] Genotype: ETH-M 24 (= DH [DAN SAN 91] × DH [PA91 × FR16]; 1,800 anthers per treatment.

[2] ES = embryo-like structures.

[3] Anther equivalent = 3500 isolated microspores in culture.

[4] Treatments followed by the same letter are not significantly different from each other (LSD: 0.05).

from the spikelets. On the other hand, the risk of contamination is higher with spikelets or tassel segments compared to excised anthers.

Specific *isolation media* have been developed to support microspore survival during isolation. During mechanical isolation, lytic compounds of the cell compartments (including the vacuoles) are released into the isolation medium. These compounds may have adverse effects on the isolated microspores. The use of a buffered isolation medium (pH = 7.0) to prevent exposure of the cells to an acidic external environment has therefore been recommended; washing the microspores immediately after isolation and isolating at low temperature are believed to reduce damage as well (Gaillard *et al.*, 1991; Vergne *et al.*, 1991).

Isolation media with various formulations of major and minor nutrients as well as organic additives have been used; the osmotic potential typically is maintained with 5–6% sucrose (Gaillard *et al.*, 1991; Genovesi & Yingling, 1994). Isolation in "regular" culture medium, i.e., YÜ-PEI based induction medium with 6% sucrose, also resulted in efficient microspore culture (Pescitelli *et al.*, 1989, 1990a).

Gaillard *et al.* (1992) reported that filter-sterilizing the isolation medium instead of autoclaving resulted in three to five-fold increased ES development. As with anther culture, heat-induced sucrose breakdown products with toxic effects may be responsible for this result (see 3.2.4). A final centrifugation step with medium containing 0.44 M sucrose has been suggested to eliminate dead microspores and less responsive, early-uninucleate microspores from the suspension (Gaillard *et al.*, 1991). In addition, beneficial effects have been obtained with a separation step at a later stage of culture: Pescitelli *et al.* (1990b) used an 88 μm sieve to retain developing microspores after seven days. This separation of developing ES from non-induced, decaying microspores led to increased microspore survival and improved ES-production.

#### 4.2.4. *Microspore plating density*

Plating density, i.e., the number of microspores per volume of medium, affects the availability of nutrients and the accumulation of metabolic by-products in the medium. The optimal plating density depends on the quality of the donor material (genotype, growth conditions of donor plants, pretreatments) as well as on the isolation and separation techniques applied. Dead microspores might release toxic substances into the culture medium and thus disturb the continued development of viable microspores (Cho & Zapata, 1990). Conversely, developing embryogenic microspores may produce and release compounds which support androgenic development in otherwise non-developing microspores. Suspensions with high frequencies of developing microspores and only a small fraction of dead microspores therefore will have a different optimal plating density than suspensions with relatively high amounts of dead microspores.

Typical plating densities in other species range from 2–5 × $10^4$ microspores per mL, e.g., 3–4 × $10^4$ per mL in rapeseed (Huang *et al.*, 1990) and 2 × $10^4$ in barley (Hoekstra *et al.*, 1993). In maize, conflicting data have been reported: Gaillard *et al.* (1991) proposed an optimal density of 6–8 × $10^4$ microspores per mL, whereas considerably lower plating densities, e.g., $10^4$ microspores per mL, have been preferred by Pescitelli *et al.* (1990a) and Afele *et al.* (1992). For barley, Hoekstra *et al.* (1993) reported a lower threshold of 5 × $10^3$ microspores per mL to observe androgenic development; at lower densities further development of the microspores was not observed. No indication of such a threshold has been found in maize.

#### 4.2.5. *Post-inoculation temperature stress*

Controversial results have been reported concerning the effects of post inoculation temperature treatments: both initial high temperatures (30–32°C) and initial cold temperatures (14°C) were reported to have beneficial effects on culture response (Coumans *et al.*, 1989; Pescitelli *et al.*, 1990a; Afele *et al.*, 1992; Genovesi & Yingling, 1994). A study that directly compared heat and cold treatments for different genotypes indicated that, as with anther culture, the post inoculation temperature treatments may act in a genotype-dependent manner (Büter, 1992).

#### 4.2.6. *Gaseous environment*

The gaseous atmosphere during induction was reported to be a critical factor in maize microspore cultures. Gaillard *et al.* (1991) obtained low ES production when microspores were cultured in individual 60 × 15 mm Petri dishes with sealed lids. In contrast, ES production greatly improved when microspores were cultured in 60 × 15 mm Petri dishes without lids and placed into a large 140 × 20 mm Petri dish together with two 60 × 15 mm Petri dishes (also without lids) containing anthers (Fig. 7). Similar effects were observed in experiments in our laboratory using different donor plant genotypes (Nägeli & Büter, unpublished).

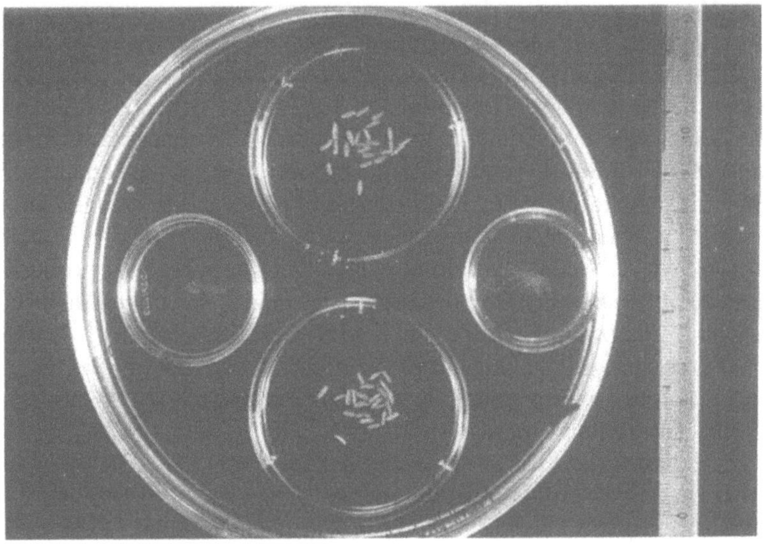

*Figure* 7. Co-culture of anthers and isolated microspores in order to produce an optimal gaseous environment for androgenic microspore development. (Note: culture dishes with anthers/microspores without lids.)

Ethylene has been found to affect androgenesis in other species; it seems that a certain level of ethylene in the culture atmosphere is required for optimal androgenic response (Biddington & Robinson, 1991; Evans & Batty, 1994). Perhaps a similar ethylene action is present in maize microspore culture.

4.2.7. *Induction media*

Induction media used for microspore cultures are generally similar to anther culture media. Typically liquid media with $N_6$ or YÜ-PEI major and minor salts are used (Pescitelli *et al.*, 1989; Büter, 1992). Fe (II) is usually applied as FeEDTA ($10^{-4}$ M). ES production also has been accomplished with a $B_5$ medium originally developed for rape seed anther culture (Coumans *et al.*, 1989). Genovesi & Yingling (1994) suggested a modified N6 basal medium characterized by a nitrate:ammonium nitrogen ratio of 6:1 ("6N1").

Various organic and inorganic compounds have been added to the induction media, e.g., activated charcoal (removed after autoclaving), vitamins, silver nitrate, L-proline, L-glutamine, L-serine, casein hydrolysate, and 2,3,5-triiodobenzoic acid (Pescitelli *et al.*, 1989; Gaillard *et al.*, 1991; Büter, 1992; Genovesi & Yingling, 1994).

Carbohydrates usually has been provided as sucrose; attempts to replace sucrose by other carbohydrates (e.g., maltose, raffinose, cellobiose, lactose, trehalose) led to decreased microspore response (Pescitelli *et al.*, 1994). Sucrose concentrations in the induction media have varied between 9 and 12%; a comparison of various sucrose concentrations (4.5, 6.0, 7.9, 9.6, 13.0, 18.0%) revealed highest ES production on media containing 7.9 or 9.6% sucrose (Pescitelli *et al.*, 1990a).

4.2.8. *Plant regeneration*

Procedures to regenerate plants from culture of isolated microspores are similar to those used in anther culture. Direct regeneration of plants and indirect regeneration after callus induction have both been accomplished with microspore cultures. In contrast to anther culture, however, no data are available on the effect of the time of ES-transfer (or transfer of multicellular, microspore-derived structures) to regeneration medium.

4.2.9. *Chromosome doubling*

Although no data have been published on the frequency of spontaneous chromosome doubling in cultures of isolated microspores, it is expected to be similar to those reported for maize anther culture. Genovesi & Yingling (1994) described a doubling step in their microspore culture procedure: colchicine (0.05%) was applied during anther preculture for 10–14 days at 10°C resulting in 25% fertile plants among the regenerants. However, no data were provided concerning the frequency of fertile plants in treatments without colchicine.

## 5. Genetics of *in vitro* androgenesis

Methodical optimization has allowed for considerable improvement of the androgenic response in maize anther and microspore culture. However, most studies have been conducted with selected, highly responsive germplasm. Screening a wide range of genotypes including commercial germplasm revealed that:

i) there exists considerable genotypic variability with regard to the androgenic response,
ii) compared to barley and wheat, only a few genotypes show a good positive response (Table 8),
iii) the level of androgenic response most frequently is unsatisfactorily low (Table 8).

In the studies cited above, typically about 1000 or fewer anthers were plated for each genotype. It would have been desirable to inoculate more anthers to increase evidence. In a study where 6000–8000 anthers/genotype were plated, quite different results were obtained: 50% of the genotypes in a set of 100 genetically unrelated, commercial lines were found to produce ES. For 30% of the tested genotypes, plants could be recovered (Beckert, personal communication).

*Table* 8. Comparison of androgenic response in barley, wheat and maize

| Species | Genotypes tested [No.] | Genot. with response[1] [%] | Genot. with good response[2] [%] | Genot. with plant regeneration [%] | Reference |
|---|---|---|---|---|---|
| Barley | 11 | 100 | 100 | 100 | Logue *et al.*, 1993 |
| | 16 | 100 | 100 | 100 | Hou *et al.*, 1994 |
| Wheat | 31 | 97 | 77 | 77 | Masojc *et al.*, 1993 |
| | 60 | 98 | 23 | 35 | Orlov *et al.*, 1993 |
| Maize | 40 | 60 | 3 | 19 | Petolino & Jones, 1986 |
| | 55 | 47 | 4 | 7 | Hongchang *et al.*, 1991 |

[1] Response = formation of at least one embryo-like structure.
[2] Good response = formation of at least 10 ES/100 anthers.

This genotypic variation indicates that genetic components are significant for the androgenic response. First genetic analyses directed to study the inheritance of *in vitro* androgenesis suggested that only a few genes are involved in the determination of the androgenic response; furthermore it was concluded from these analyses that the induction process leading to the formation of ES and the subsequent regeneration of plants are not correlated and determined by different genes (Barloy *et al.*, 1989; Afele & Kannenberg, 1990). It also was demonstrated that general combining ability is highly significant, i.e., the average response of a specific parental line can be used to predict the response level that can be expected in a given cross (Petolino & Thompson, 1987). This pattern is usually characteristic for additive gene effects and quantitatively inherited traits in maize. Other heritability studies confirmed that additive genetic variance for androgenic response (ES and plant formation in anther culture) was greater than non-additive genetic variance (Barloy, 1990; Afele & Kannenberg, 1990). Since specific combining ability also was found to be important, it has been assumed that dominance effects are also relevant (Petolino & Thompson, 1987).

The fact that genetic components are involved in the androgenic response implies that plants regenerated by anther or microspore culture contain at least some of these favorable genes; thus, anther and microspore culture can be applied as selective processes for the accumulation of genes favoring improved androgenic response. Corresponding experiments showed that a single cycle of selection resulted in more than a sixfold increase in the androgenic response as indicated by the frequency of ES production (Petolino, 1989).

If methodical optimization should fail to overcome the genotypic limitations in maize androgenesis, an alternate approach to extend the spectrum of responsive germplasm, e.g., to commercially important materials, might be to transfer the corresponding genes from highly to less responsive germplasm. In fact, significant progress has been made by intermating and subsequent selection for highly-responsive genotypes (Barloy *et al.*, 1989; Büter *et al.*, 1994). However, selecting for androgenic aptitude normally involves labor intensive, anther or microspore culture-based screening procedures. Obviously a marker-based selection would greatly simplify the identification of responsive and non-responsive germplasm.

Recently, the availability of genetic linkage maps containing many markers which cover the entire genome at close intervals has raised the possibility of marker-aided selection. Genetic markers such as RFLPs allow the identification of single gene components which contribute to complex, typically quantitatively inherited traits (quantitative trait loci or QTL). By correlating the segregation of a given trait, e.g., androgenic response, with a set of RFLP markers with known chromosomal positions, important loci affecting this trait can be tracked.

Using molecular approaches valuable information concerning the genetic determination for androgenic response in maize has been elaborated. The first report on a QTL analysis used $S_1$ families derived from a cross between a nonresponsive ("B 73") and a highly responsive genotype [$S_4$ line of "139/39", a genotype derived from a cross between two anther-derived DH plants of the hybrid, (H99 × FR16) × PA91; Cowen *et al.*, 1992]. A total of 98 families was subjected to RFLP analysis. RFLP data were correlated to ES formation; plant regeneration was not considered. It was concluded that the anther culture response observed in the highly responding parent was conditioned by two major recessive genes located on chromosomes 3 and 9. These genes were epistatic. Two minor genes located on chromosomes 1 and 10 also seemed to be involved. In summary, these four chromosomal regions were found to explain 57% of the variability among the families.

A second study used anther culture-derived, totipotent callus lines of three $F_1$ hybrids with a pedigree similar to that of the responsive parent used in the first study (Wan *et al.*, 1992). Formation of ES and the production of regenerable calli were used as criteria to quantify the level of the androgenic response. In agreement with Cowen's study, two regions with favorable alleles were identified on chromosomes 1 and 3. Additionally four regions on chromosomes 2 (2 regions), 6 and 8 were found to be relevant, either for ES or regenerable callus formation, while no favorable alleles were detected on chromosomes 9 and 10.

The most comprehensive study published thus far used three different single cross hybrids including highly responsive and nonresponsive parental lines (Murigneux *et al.*, 1994). More than 200 DH lines were generated from these three crosses by anther culture. The percentage of androgenic anthers ("anther culturability"), the number of ES produced by 100 anthers, the number of regenerated plants by 100 ES and the number of regenerated plants by 100 anthers were determined and used as criteria to describe the level of the androgenic response. The correlations of the anther culture and RFLP data led to following conclusions:

i) anther culturability was highly heritable (mean heritability of 0.90 and 0.78 in two different crosses);
ii) anther culturability was the most useful early indicator for the final yield of regenerated plants. Generally, the potential to produce ES seemed to be more important for the final yield of regenerated plants than the regeneration potential of the ES;
iii) estimations of allele frequencies revealed that selection for positive alleles apparently does not take place during anther culture. If *in vitro* androgenesis were completely under gametophytic control, such selection would have been expected. Thus it appeared that sporophytic components are involved. Similar results have been obtained by Cowen *et al.* (1992);
iv) distributions and ranges of variation of the recombinant lines indicated that a small number (5–6) of genes or groups of genes led to the different anther culture response in the responsive and non-responsive parental lines;
v) from a cross between a responsive (DH 7) and non-responsive (A 188) line, transgressive lines were obtained which showed a higher anther culture response than the responsive parent. Therefore it was assumed that regions with positive alleles may be present even in non-responsive germplasm. Inter-mating two non-responsive lines theoretically could lead to the formation of a responsive hybrid, if different favorable alleles were present in the parental lines and acted complementarily;
vi) the map positions of the QTLs were not conserved among the three crosses; in addition, favorable alleles were identified in chromosomal regions which were different from those regions described in earlier studies based on genetically unrelated germplasm.

In conclusion, *in vitro* androgenesis in maize seems to be determined by a complex genetic system with various QTLs conditioning specific steps during the androgenic process. Potentially only genotypes equipped to perform each of these steps will give a positive response (Murigneux *et al.*, 1994). It is obvious from these results that further studies including various responsive genotypes and detailed molecular analyses of the regions associated with the androgenic response will be needed for a better understanding of the events which finally result in ES and plant formation. This may allow for the practical application of these QTLs, i.e., the purposeful construction of highly responsive genotypes.

**6. Homogeneity, segregation ratio and agronomic performance of doubled haploid lines**

As with conventionally derived inbred lines, homogeneity represents a prerequisite for the use of DH lines in maize breeding. Homogeneity of DH progenies has been investigated by the use of molecular markers (RFLPs) and in field trials by the evaluation of morphological traits, e.g., plant height, number of leaves, etc. (Bentolila *et al.*, 1992; Murigneux *et al.*, 1993b). DH lines used for the homogeneity studies were produced by direct regeneration through anther culture, i.e., no callus stage occurred before plant regeneration (see 3.1). Both molecular and field studies confirmed that the level of homogeneity in DH progenies was similar to that obtained in corresponding lines produced conventionally. Thus, anther culture, relying on a direct regeneration procedure, was a reliable method for the production of homozygous lines (Fig. 8).

For breeding purposes, it is essential that *in vitro* produced DH plants represent a random array of the microspore population, i.e., that no gametic selection occurs during *in vitro* androgenesis which would lead to a distorted segregation. Investigations into this subject were carried out by comparison of the RFLP marker segregation in DH and $F_2$ (Bentolila *et al.*, 1992) or single seed descent (SSD) lines (Murigneux *et al.*, 1993a) derived from the same cross. Both studies revealed a partial gametic selection in the DH lines. The extent of the segregation distortion seemed to be affected by the difference in anther culture capacity of the two parental lines, i.e., distortion was relatively frequent in crosses involving lines with different anther culture capacity. It has been concluded from these results that *in vitro* androgenesis in breeding programs should be applied to crosses of genotypes with similar androgenic potential, thus avoiding the risk of reduced genetic variability compared to conventional procedures.

Both studies cited here used similar germplasm; therefore, additional studies into this issue using responsive germplasm with a different genetic background certainly would be useful; however, distorted segregation has also been found in other cereal species, e.g., barley (Zivy *et al.*, 1992; Kintzios *et al.*, 1994).

Studies on the agronomic performance of DH lines and comparable conventionally derived lines [SSD and pedigree selected (PS) lines] revealed relatively little difference among them. Petolino (1989) investigated the hybrid performance of DH lines by crossing DH, SSD and PS lines to a common tester: no significant differences were detected among the lines obtained by the different methods. Murigneux *et al.* (1993a) compared DH and SSD lines ($F_6$) for the distribution of seven agromorphological traits: similarly, no major differences were observed except for two traits related to leaf characteristics.

*Figure* 8. From *in vitro* plant to DH line: plant transfer to soil (above left); fertile DH-plant (above right); seeds from selfed plant (below left); field evaluation of DH-line (below right).

## 7. Haploid cells as targets for genetic transformation

Haploid cells are desirable targets for the delivery of foreign genes, since they would allow the integration into a completely homozygous plant (if chromosome doubling were successful). Delivery into a homozygous genome would omit the recovery of transformants with lethal mutants. As a unicellular system, isolated microspores as a transformation target would avoid the formation of chimeric regenerants. Moreover, the risk of inducing *in vitro* culture-related somaclonal variation is relatively low, since the time in culture is short compared to other target tissue, e.g., regenerable suspensions or embryogenic callus cultures.

Due to these obvious advantages of the microspore system, many efforts have been made to develop gene transfer procedures suitable for microspore transformation. To date, the biolistic approach appears to be most promising since it allowed successful stable transformation in barley (Jähne *et al.*, 1994) and tobacco (Stöger *et al.*, 1995). In maize, microspore transformation by particle bombardment has been attempted, but no stable transformation has been accomplished (Beckert, personal communication; Petolino, personal communication). However, by GUS-expression it was shown that cultured microspores can take up and express foreign DNA after biolistic transformation (Jardinaud *et al.*, 1995). Generally, a major problem with maize microspore transformation appears to be the insufficient frequency of embryogenic microspores during culture initiation. Innumerable microspores need to be bombarded in order to transform one embryogenic microspore which subsequently manages to regenerate into a fertile DH plant. This approach may eventually lead to stable transformed, homozygous plants but it appears that the microspore system needs to be more efficient before it can be used for gene transfer into maize.

Attempts to circumvent the problem of low plating efficiency have made use of microspore-derived ES at early developmental stages as targets. In a preliminary study including various stages of proembryo development (100–400 μm in size) Gaillard *et al.* (1992) injected lucifer yellow dye in order to find the most suitable stage for DNA delivery by microinjection. Their results suggested multinuclear structures, i.e., proembryos 120–140 μm in size immediately after rupture of the exine, but before cell wall formation, should be used for DNA microinjection. Pescitelli *et al.* (1990b) used proembryos separated from the microspore suspension 2, 4, 7 and 21 days after culture initiation. DNA was transferred by particle bombardment, but no stable transformation was accomplished.

Following a different approach, Wan *et al.* (1995) used anther culture-derived type I callus after chromosome doubling for microprojectile bombardment. A total of twelve transgenic callus lines was obtained, each of them giving rise to multiple fertile, stably transformed plants.

Electroporation and polyethylene glycol-mediated DNA transfer have been suggested as a means of delivering DNA into maize microspores;

both methods proved to be effective for delivering free DNA into maize microspores (Fennel & Hauptmann, 1992). In contrast to particle bombardment and microinjection, these methods allow for the rapid treatment of a large number of individual receptive cells. Using the electroporation method Sukhapinda *et al.* (1993) were able to regenerate stably transformed maize from protoplasts isolated from microspore-derived suspension cultures.

## 8. Conclusions and prospects

Research on the methodical optimization of *in vitro* maize androgenesis began in 1975 (Anonymous, 1975). No one could deny that 20 years of research has led to remarkable improvements of the androgenic response in maize anther and microspore culture. However, with regard to practical use, applicability to a broad spectrum of genotypes including elite germplasm, will be essential. Even in many elite lines, a certain degree of androgenic potential appears to be present which can be enhanced and exploited. Nevertheless, for breeders who intend to use anther or microspore culture to derive DH lines it is advisable to accumulate the androgenic potential in their genetic material in order to realize reliable and cost-efficient DH production. In order to facilitate this accumulation, a marker-aided selection is highly desirable.

Optimists believe that further improvements of donor plant growth conditions and *in vitro* culture methods eventually may remove or greatly reduce present genotypic limitations. Although history does not allow for too much optimism, further research into the genes and gene products which affect the androgenic response may one day allow production of DH plants from previously "non-responsive" genotypes.

## 9. References

Afele, J.Y. & L.W. Kannenberg (1990) Genetic studies of corn (*Zea mays* L.) anther culture response. Theor. Appl. Genet. 80: 459–464.

Afele J.C., L.W. Kannenberg, R. Keats, S. Sohota & E.B. Swanson (1992): Increased induction of microspore embryos following manipulation of donor plant environment and culture temperature in corn (*Zea mays* L.). Plant Cell Tiss. Organ Cult. 28: 87–90.

Anonymous 401 Research Group (1975) Primary study on induction of pollen plants of *Zea mays*. Acta Genet. Sin. 2: 143.

Ao, G.M., S.X. Zhao & G.H. Li (1982) *In vitro* induction of haploid plantlets from unpollinated ovaries of corn (*Zea mays* L.). Acta Genet. Sin. 9: 281–283.

Barloy, D. (1990) Étude de facteurs environnementaux et génétiques de l'androgenèse *in vitro* chez le maïs. Caractérisation de lignées haploïdes doublées. Thèse 232, Université de Clermont II.

Barloy, D. & M. Beckert (1993) Improvement of the regeneration ability of androgenetic embryos by early anther transfer in maize. Plant Cell Tiss. Organ Cult. 33: 45–50.

Barloy, D., L. Denis & M. Beckert (1989) Comparison of the aptitude of anther culture in some androgenetic doubled haploid maize lines. Maydica 34: 303–308.

Barnabas, B., P.F. Fransz & J.H. Schel (1987) Ultrastructure studies on pollen embryogenesis in maize (*Zea mays* L.). Plant Cell Rep. 6: 212–215.

Bedinger, P.A. & M.D. Edgerton (1990) Developmental staging of maize microspores reveals a transition in developing microspore proteins. Plant Physiol. 92: 474–479.

Bentolila, S., T. Hardy, C. Guitton & G. Freyssinet (1992) Comparative genetic analyses of $F_2$ plants and anther culture derived plants of maize. Genome 35: 575–582.

Biddington, N.L. & H.T. Robinson (1991) Ethylene production during anther culture of Brussel sprouts (*Brassica oleracea* var. *gemmifera*) and its relationship with factors that affect embryo production. Plant Cell Tiss. Organ Cult. 25: 169–177.

Brettel, R.I., E. Thomas & W. Wernicke (1981) Production of haploid maize plants by anther culture. Maydica 26: 101–111.

Büter, B. (1992) Einflussfaktoren der *in vitro* Haploidinduktion bei *Zea mays*. Ph.D. Thesis No. 9725, Swiss Federal Institute of Technology (ETH), Zürich, Switzerland.

Büter, B., C. Jumpatong, K. Berger-Büter, S. Saisingtong, J.E. Schmid, R. Thiraporn & P. Stamp (1994) Application of the anther culture technique in a corn breeding project in Thailand. Abstracts VIIIth International Congress of Plant Tissue and Cell Culture, Firenze, June 12–17, p. 85.

Büter, B., S.M. Pescitelli, K. Berger, J.E. Schmid & P. Stamp (1993) Autoclaved and filter-sterilized liquid media in maize anther culture: significance of activated charcoal. Plant Cell Rep. 13: 79–82.

Büter, B., S. Saisingtong, J.E. Schmid, M. Nägeli & P. Stamp (1995) Improved methods for the *in vitro* production of doubled haploids in maize. In: U. Wienand (Ed.), MEN, 2-Maize European Network Meeting, Hamburg, 29–31 May 1995, in press.

Büter, B., J.E. Schmid & P. Stamp (1990) Effect of gametocides on the induction of androgenesis in wheat (*Triticum aestivum*) and maize (*Zea mays* L.). Abstracts VIIth International Congress on Plant Tissue and Cell Culture, Amsterdam, June 24–29, p. 184.

Büter, B., J.E. Schmid & P. Stamp (1991) Effects of L-proline and post-plating temperature treatment on maize (*Zea mays* L.) anther culture. Plant Cell Rep. 10: 325–328.

Cai, Q., I. Szarejko, K. Polok & M. Maluszynski (1992) The effect of sugars and growth regulators on embryoid formation and plant regeneration from barley anther culture. Plant Breed. 109: 218–226.

Cho, M.S. & F.J. Zapata (1990) Plant regeneration from isolated microspores of Indica rice. Plant Cell Physiol. 31: 881–885.

Chu, C.C. (1981) The $N_6$-medium and its applications to anther culture of cereal crops. Proc. Symp. Plant Tissue Culture, Beijing 1978, pp. 43–50. Pitman, Boston.

Chu, C.C., R.D. Hill & A.L. Brule-Babel (1990) High frequency of pollen embryoid formation and plant regeneration in *Triticum aestivum* L. on monosaccharide containing media. Plant Sci. 66: 255–262.

Coe, E.H. (1959) A line with a high haploid frequency. Am. Nat. 93: 381–382.

Coumans, M.P., S. Sohota & E.B. Swanson (1989) Plant development from isolated microspores of *Zea mays* L. Plant Cell Rep. 7: 618–621.

Cowen, N.M., C.D. Johnson, K. Armstrong, M. Miller, A. Woosley, S. Pescitelli, M. Skokut, S. Belmar & J.F. Petolino (1992) Mapping genes conditioning *in vitro* androgenesis in maize using RFLP analysis. Theor. Appl. Genet. 84: 720–724.

Custers, J.B., J.H. Cordewener, Y. Nöllen, H.J. Dons & M.M. van Lockeren-Campagne (1994) Temperature controls both the gametophytic and sporophytic development in microspore cultures of *Brassica napus*. Plant Cell Rep. 13: 267–271.

Deimling, S., T. Flehinghaus, I. Schneider & H.H. Geiger (1990) Influence of post-plating temperature and carbohydrate source in maize and rye anther culture. Abstracts VIIth International Congress on Plant Tissue and Cell Culture, Amsterdam, June 24–29, p. 195.

Dieu, P. & M. Beckert (1986) Further studies of androgenetic embryo production and plant regeneration from *in vitro* cultured anthers of maize (*Zea mays* L.). Maydica 31: 245–259.

Evans, J.M. & N.P. Batty (1994) Ethylene precursors and antagonists increase embryogenesis of *Hordeum vulgare* L. anther culture. Plant Cell Rep. 13: 676–678.

FAO [Food and Agriculture Organization of the United Nations] (Ed.) (1994) FAO Yearbook Production, Vol. 47, 1993, FAO Statistics Series No. 117.

Fennel, A. & R. Hauptmann (1992) Electroporation and PEG delivery of DNA into maize microspores. Plant Cell Rep. 11: 567–570.

Finnie, S.J., W. Powell & A.F. Dyer (1989) The effect of carbohydrate composition and concentration on anther culture response in barley (*Hordeum vulgare* L.). Plant Breed. 103: 110–118.

Flehinghaus, T., S. Deimling & H.H. Geiger (1991) Methodical improvements in rye anther culture. Plant Cell Rep. 10: 397–400.

Gaillard, A., E. Matthys-Rochon & C. Dumas (1992) Selection of microspore derived embryogenic structures in maize related to transformation potential by microinjection. Bot. Acta 105: 313–318.

Gaillard, A., P. Vergne & M. Beckert (1991) Optimization of maize microspore isolation and culture conditions for reliable plant regeneration. Plant Cell Rep. 10: 55–58.

Genovesi, A.D. (1990) Maize (*Zea mays* L.): *In vitro* production of haploids. In: Y.P.S. Bajaj (Ed.), Biotechnology in Agriculture and Forestry 12: Haploids in Crop Improvement I, pp. 176–203. Springer-Verlag, Berlin.

Genovesi, A.D. & G.B. Collins (1982) *In vitro* production of haploid plants of corn via anther culture. Crop Sci. 22: 1137–1144.

Genovesi, A.D. & R.A. Yingling (1990) Isolated microspore culture of *Zea mays* L. Abstracts VIIth International Congress on Plant Tissue and Cell Culture, Amsterdam, June 24–29, p. 196.

Genovesi, A.D. & R.A. Yingling (1994) Isolated microspore and anther culture of corn. United States Patent, Patent Number 5,322,789, June 21, 1994, United States Patent and Trademark Office, Washington DC, USA.

Henry, Y. & J. de Buyser (1981) Float culture of wheat anthers. Theor. Appl. Genet. 60: 77–79.

Henry, Y. & J. de Buyser (1990) Wheat anther culture: agronomic performance of doubled haploid lines and the release of a new variety "Florin". In: Y.P.S. Bajaj (Ed.), Biotechnology in Agriculture and Forestry 12: Haploids in Crop Improvement I, pp. 285–352. Springer-Verlag, Berlin.

Hoekstra, S., M.H. van Zijderveld, F. Heidekamp & F. van der Mark (1993) Microspore culture of *Hordeum vulgare* L.: the influence of density and osmolality. Plant Cell Rep. 12: 661–665.

Hongchang, M., G.H. Liang & C.E. Wassom (1991) Effects of growth regulators and genotypes on callus and embryoid induction from maize anther culture. Plant Breed. 106: 47–52.

Hou, L., E. Ullrich, A. Kleinhofs & C.M. Stiff (1993) Improvement of anther culture methods for doubled haploid production in barley breeding. Plant Cell Rep. 12: 334–338.

Hou, L., S.E. Ullrich & A. Kleinhofs (1994) Inheritance of anther culture traits in barley. Crop Sci. 34: 1243–1247.

Huang, B., S. Bird, R. Kemble, D. Simmonds, W. Keller & B. Miki (1990) Effects of culture density, conditioned medium and feeder cultures on microspore embryogenesis in *Brassica napus* L. cv. Topas. Plant Cell Rep. 8: 594–597.

Hunter, C.P. (1987) European Patent Application by Shell International Research Maatschappij B.V. No. 87200773.7.

Jähne, A., D. Becker, R. Brettschneider & H. Lörz (1994) Regeneration of transgenic, microspore-derived, fertile barley. Theor. Appl. Genet. 89: 525–533.

Jardinaud, M.F., A. Souvre, G. Alibert & M. Beckert (1995) *uidA* Gene transfer and expression in maize microspores using the biolistic method. Protoplasma 187: 138–143.

Kermicle, J.L. (1969) Androgenesis conditioned by a mutation in maize. Science 166: 1422–1424.

Kintzios, S., R.M. Islam & G. Fischbeck (1994) Distorted segregation for mildew resistance in doubled haploid lines of spring barley. Plant Breed. 112: 248–251.

Ku, M.K., W.C. Cheng, L.C. Juo, Y.L. Kuan, H.P. An & C.H. Huang (1981) Induction factors and morpho-cytological characteristics of pollen derived plants in maize (*Zea mays* L.). In: Proc Symp Plant Tissue Culture 1978, pp. 35–42. Science Press, Peking.

Kuo, C.S., W. Lu & Y.L. Kui (1986) Corn (*Zea mays* L.) Production of pure lines through anther culture. In: Y.P.S. Bajaj (Ed.), Biotechnology in Agriculture and Forestry, Vol. 2. Crops I, pp. 152–164. Springer-Verlag, Berlin.

Lashermes, P. & M. Beckert (1988) Genetic control of maternal haploidy in maize (*Zea mays* L.) and selection of haploid inducing lines. Theor. Appl. Genet. 76: 405–410.

Logue, S.J., L.C. Giles & D.H. Sparrow (1993) Genotype and environment strongly influence barley anther culture response using Australian genotypes. Aust. J. Bot. 41: 227–236.

Lyne, R.L., R.I. Bennet & C.P. Hunter (1986) Embryoid and plant production from cultured barley anthers. In: L.A. Withers & P.G. Anderson (Eds.), Plant Tissue Culture and its Agricultural Application, pp. 405–411. Butterworths, Guildford.

MacDonald, M.V. (1992) Donor plant growth factors affecting anther culture of maize and sweet corn (*Zea mays*). Ann. Bot. 70: 357–363.

Masojc, P., O.M. Lukow, M.C. Kenzie & N.K. Howes (1993) Responsiveness to anther culture in cultivars and $F_1$ crosses of spring wheat. Can. J. Plant Sci. 73: 777–783.

Miao, S.H., C.W. Kuo, Y.L. Kwei, A.T. Sun, S.Y. Ku, W.L. Lu, Y.Y. Wang, M.L. Chen, M.K. Wu & L. Hang (1978) Induction of pollen plants of maize and observations on their progeny. In: Proceedings of Symposium on Plant Tissue Culture, pp. 23–33. Science Press, Peking.

Murigneux, A., S. Baud & M. Beckert (1993a) Molecular and morphological evaluation of doubled haploid lines in maize. 2. Comparison with single-seed-descent lines. Theor. Appl. Genet. 87: 278–287.

Murigneux, A., D. Barloy, P. Leroy & M. Beckert (1993b) Molecular and morphological evaluation of doubled haploid lines in maize. 1. Homogeneity within DH lines. Theor. Appl. Genet. 86: 837–842.

Murigneux, A., S. Bentolila, T. Hardy, S. Baud, C. Guitton, H. Jullien, S. Ben Tahar, G. Freyssinet & M. Beckert (1994) Genotypic variation of quantitative trait loci controlling *in vitro* androgenesis in maize. Genome 37: 970–976.

Nitsch, C. (1977) Culture of isolated microspores. In: J. Reinert & Y.P.S. Bajaj (Eds.), Applied and Fundamental Aspects of Plant Cell, Tissue, and Organ Culture, pp. 268–278. Springer-Verlag, Berlin.

Nitsch, C., S. Andersen, M. Godard, M.G. Neuffer & W.F. Sheridan (1982) Production of haploid plants of *Zea mays* and *Pennisetum* through androgenesis. In: E.D. Earle & Y. Demarly (Eds.), Variability in Plants Regenerated from Tissue Culture, pp. 66–91. Praeger, New York.

Orlov, P.A., E.B. Mavrishcheva & A.N. Palilova (1993) Estimation of the response to anther culturing in 60 genotypes of different wheat species. Plant Breed. 111: 339–342.

Orshinsky, B.R., L.J. McGregor, G. Johnson, P. Hucl & K.K. Kartha (1990) Improved embryoid induction and green shoot regeneration from wheat anthers cultured on medium with maltose. Plant Cell Rep. 9: 365–369.

Pace, G.M., J.N. Reed, L.C. Ho & J.W. Fahey (1987) Anther culture of maize and the visualization of embryogenic microspores by fluorescent microscopy. Theor. Appl. Genet. 73: 863–869.

Pauk, J. (1985) Production of haploid plants of maize (*Zea mays* L.) through androgenesis. Cereal Res. Commun. 13: 47–53.

Pescitelli, S.M. & J.F. Petolino (1988) Microspore development in cultured maize anthers. Plant Cell Rep. 7: 441–444.

Pescitelli, S.M., C.D. Johnson & J.F. Petolino (1990a) Isolated microspore culture of maize: effects of isolation technique, reduced temperature, and sucrose level. Plant Cell Rep. 8: 628–631.

Pescitelli, S.M., C.D. Johnson & J.F. Petolino (1990b) Isolated microspore culture of maize.

Abstracts VIIth International Congress on Plant Tissue and Cell Culture, Amsterdam, June 24–29, p. 189.

Pescitelli, S.M., C.D. Johnson & J.F. Petolino (1994) Isolated microspore culture of maize. In: Y.P.S. Bajaj (Ed.), Biotechnology in Agriculture and Forestry, Vol. 25, Maize, pp. 186–200. Springer-Verlag, Berlin.

Pescitelli, S.M., J.C. Mitchell, A.M. Jones, D.R. Pareddy & J.F. Petolino (1989) High frequency androgenesis from isolated microspores of maize. Plant Cell Rep. 7: 673–676.

Petolino, J.F. (1989) Use of anther culture and related procedures for corn improvement. In: Proceedings of the Forty-Fourth Annual Corn and Sorghum Industry Research Conference, American Seed Trade Association, Washington, DC.

Petolino, J.F. & A.D. Genovesi (1994) Anther and microspore culture. In: M. Freeling & V. Walbot (Eds.), The Maize Handbook, pp. 701–704. Springer-Verlag, New York.

Petolino, J.F. & A.M. Jones (1986) Anther culture of elite genotypes of maize. Crop Sci. 26: 1072–1074.

Petolino, J.F. & S.A. Thompson (1987) Genetic analysis of anther culture response in maize. Theor. Appl. Genet. 74: 284–286.

Pretova, A., N. de Ruijter, A. van Lammeren & J.H. Schel (1993) Structural observations during androgenic microspore culture of the 4c1 genotype of *Zea mays* L. Euphytica 65: 61–69.

Saisingtong, S., J.E. Schmid, P. Stamp & B. Büter (1996) Colchicine-mediated chromosome doubling during anther culture of maize (*Zea mays* L.). Theoret. Appl. Genet. 92: 1017–1023.

Santos, M., N. Boget & J.M. Tome (1995) Endogenous polyamine content during *in vivo* maturation and *in vitro* culture of maize pollen. Plant Growth Reg. 16: 19–26.

Schmid, J.E. & E.R. Keller (1986) Effect of a gametocide on the induction of haploids in *Triticum aestivum*. In: W. Horn, C.J. Jensen, W. Odenbach & O. Schieder (Eds.), Proc. Int. Symp. Eucarpia, Sept. 8–13, Berlin (West), pp. 347–349. De Gruyter, Berlin.

Stöger, E., C. Ink, M. Pfosser & E. Heberle-Bors (1995) Plant transformation by particle bombardment of embryogenic pollen. Plant Cell Rep. 14: 273–278.

Sukhapinda, K., M.E. Kozuch, B. Rubin-Wilson, W.M. Ainley & D.J. Merlo (1993) Transformation of maize (*Zea mays* L.) protoplasts and regeneration of haploid transgenic plants. Plant Cell Rep. 13: 63–68.

Theander, O. & D.A. Nelson (1988) Aqueous, high-temperature transformation of carbohydrates relative to utilization of biomass. Adv. Carbohydrate Chem. Biochem. 46: 273–326.

Ting, Y.C., M. Yu & Z.Z. Wan (1981) Improved anther culture of maize (*Zea mays*). Plant Sci. Lett. 23: 139–145.

Truong-Andre, I. & Y. Demarly (1984) Obtaining plants by *in vitro* culture of unfertilized maize ovaries (*Zea mays* L.) and preliminary studies on the progeny of a gynogenetic plant. Z. Pflanzenzüchtg. 92: 309–320.

Tsay, H.S., S.H. Miao & J.M. Widholm (1986) Factors affecting haploid plant regeneration from maize anther culture. J. Plant Physiol. 126: 33–40.

Vergne, P., A. Gaillard & M. Beckert (1991) Isolation of viable microspores and immature pollen grains from cereal inflorescences. In: I. Negrutiu & G. Gharti-Chhetri (Eds.), A Laboratory Guide for Cellular and Molecular Plant Biology, pp. 75–87. Birkhäuser Verlag, Basel/Boston/Berlin.

Vergne, P., F. Riccardi, M. Beckert & C. Dumas (1993) Identification of a 32-kDa anther marker protein for androgenic response in maize, *Zea mays* L. Theor. Appl. Genet. 86: 843–850.

Wan, Y., D.R. Duncan, A.L. Rayburn, J.F. Petolino & J.M. Widholm (1991) The use of antimicrotubule herbicides for the production cf doubled haploid plants from anther-derived maize callus. Theor. Appl. Genet. 81: 205–211.

Wan, Y., J.F. Petolino & J.M. Widholm (1989) Efficient production of doubled haploid plants through colchicine treatment of anther-derived maize callus. Theor. Appl. Genet. 77: 889–892.

Wan, Y., T.R. Rocheford & J.M. Widholm (1992) RFLP analysis to identify putative chromosomal regions involved in the anther culture response and callus formation. Theor. Appl. Genet. 85: 360–365.

Wan, Y. & J.M. Widholm (1992) Formation of multiple embryo-like structures from single microspores during maize anther culture. Plant Cell Rep. 11: 529–531.

Wan, Y. & J.M. Widholm (1993) Anther culture of maize. Plant Breeding Rev. 11: 199–224.

Wan, Y., J.M. Widholm & P.G. Lemaux (1995) Type I callus as a bombardment target for generating fertile transgenic maize (*Zea mays* L.). Planta 196: 7–14.

Weatherhead, M.A., J. Burdon & G.G. Henshaw (1979) Effects of activated charcoal as an additive to plant tissue culture media: part 2. Z. Pflanzenphysiol. 94: 399–405.

Ziauddin, A. & K.J. Kasha (1990) Genetic stability of haploid cell cultures. In: Y.P.S. Bajaj (Ed.), Biotechnology in Agriculture and Forestry, Vol. 12, Haploids in Crop Improvement I, pp. 83–98. Springer-Verlag, Berlin.

Zivy, M., P. Devaux, J. Blaisonneau, R. Jean & H. Thiellement (1992) Segregation distortion and linkage studies in microspore-derived double haploid lines of *Hordeum vulgare* L. Theor. Appl. Genet. 83: 919–924.

# 3. *In vitro* induced haploids in wheat

HAN HU

**Contents**



## 1. Introduction

Haploids have attracted great interest of geneticists, plant embryologists, physiologists and breeders since the first discovery of haploid plants in *Datura stramonium* (Blakeslee *et al.*, 1922). In the beginning haploid plants were regarded as a special biological phenomenon. The occurrence of haploids was reported in 71 species of angiosperms belonging to 39 genera and 14 families up to early 1960s (Kimber & Riley, 1963). However, the research and utilization of haploids were limited then by their low frequency. In the last three decades many techniques especially *in vitro* culture have been developed to induce large numbers of haploid plants. For instance, by anther culture, haploids were induced in 247 species of angiosperms belonging to 88 genera of 34 families (Maheshwari *et al.*, 1983). In addition, chromosome elimination (the *bulbosum* technique) and *in vitro* culture of unfertilized ovaries and ovules have also proved to be efficient means of haploid induction in some plant species. This progress paved the way for studying and utilizing haploids in higher plants including wheat.

*S.M. Jain, S.K. Sopory & R.E. Veilleux (eds.),* In Vitro *Haploid Production in Higher Plants, Vol. 4,* 73–97.

The application of haploids in genetic studies and breeding is based on the following genetic characters of haploids.

Firstly, homozygous diploids can be obtained in one generation via chromosome doubling of the haploid plants, whereas six or seven generations are required for the production of homozygous diploids via conventional breeding methods. Hence haploids can be used to shorten the breeding cycle, save time, space and labor. Secondly, recessive characters are easily expressed in haploids since there is no such phenomenon as recessive genes hidden by dominant ones. This is particularly valuable in mutation breeding and mutant genetic research. Moreover, selection efficiency is enhanced as it is made at strain (population) level instead of individual level. Thirdly, gametic genotypes are fully expressed at plant level in haploids. Gametic types are not lost and multitudinous plants provide a wide spectrum for selection in breeding practice as well as facilitate genetic analysis of recombinant gametes. Fourthly, about 90% haploids and homozygous diploids are produced via anther culture together with 10% aneuploids in wheat (*Triticum aestivum*) (Hu *et al.*, 1978, 1980, 1985), maize (*Zea mays*) (Gu, 1986), and rice (*Oryza sativa*) (Chen & Li, 1978), etc. Aneuploids may be used in genetic analysis of chromosomal transfer, thus creating new genetic resources for genetic studies and breeding practice.

The above listed genetic characters of haploids have made it possible to investigate somaclonal variation and parasexuality in higher plants which is an important but difficult field in the studies of higher plants.

## 2. Origin of haploids

There are two different generations in the life cycle of higher plants: namely the asexual generation is producing spores and the sexual phase is producing gametes. The sporophyte in the asexual generation is diploid with two chromosome sets derived from both parents. Before formation of the spores through meiosis, the zygotic (diploid) chromosome number is reduced to the gametic (haploid) number, characteristic for the gametes or the haploid phase of the life cycle. That means, haploids of higher plants are sporophytes with gametic chromosomes.

Figure 1 shows the origin of haploid sporophytes of higher plants. In nature the haploids of higher plants are produced via abnormal fertilization, therefore, it is seldom seen, and the frequency of appearance of haploids is very low. Since the discovery of the first haploid plants in beginning of 1920s, many efforts have been made to induce haploids of higher plants which can be classified roughly into two categories: *in vivo* induction of haploids by various physical, chemical or biological stimulants, and *in vitro* induction culture methods which appeared to be more efficient than *in vivo* methods.

From the life cycle point of view (Fig. 1) there are three pathways to induce haploid sporophytes of higher plants.

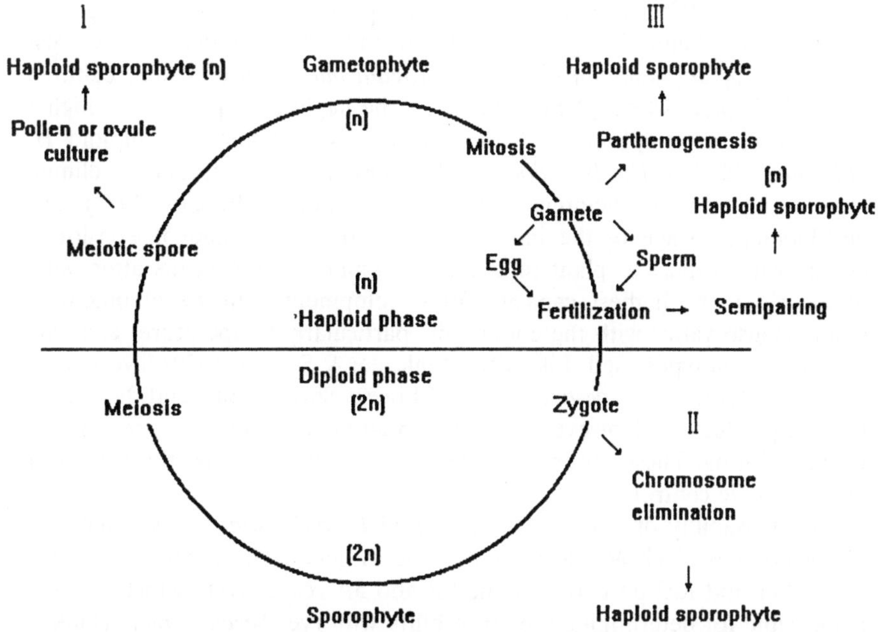


*Figure* 1. Life cycle of higher plants.

I. Haploid sporophyte originating from meiotic spore by means of pollen (anther) or/and ovary and ovule culture;
II. Haploid sporophyte originating from zygote due to chromosome elimination; and
III. Haploid sporophyte originating from male and female gametes, their precursor cells and gametophytes by isolation and manipulation of reproductive cells and protoplasts (Yang & Zhou, 1992).

By application of the final pathway it is most difficult to obtain haploids so this technique cannot be applied on cereals up to now. Although this paper will not cover it, the techniques of experimental plant reproductive biology show great potential in providing new means for biotechnology and may eventually permit direct reproductive cell engineering. Meanwhile it serves to deepen our knowledge about the control of reproductive processes.

## 3. Zygotic chromosome elimination

The chromosome elimination technique originally developed for barley (Kasha & Kao, 1970) named *bulbosum* technique, was extended to wheat by Barcley in 1975, who reported high frequency of haploids through the chromosome elimination process. Hybrid embryos resulting from a cross between *Triticum aestivum* (♀) and *Hordeum bulbosum* (♂) were cultured to obtain haploid wheat plants. However, the haploid frequency was higher (53.6%) when tetraploid *H. bulbosum* was used rather than diploid *H. bulbosum* (28.2%). The *H. bulbosum* chromosomes are preferentially eliminated during embryo development from a hybrid zygote (Barcley, 1975). The developmental stage of the immature embryos at the time of excision is critical with respect to plant regeneration; embryos at 2 weeks after pollination show the highest capacity for development. The development of embryos also varies with the conditions, particularly temperature, and with the wheat genotypes used. Likewise, Zenkteler & Straub (1979) investigated three genotypes of *T. aestivum*, i.e., Chinese Spring, Bali and Janus, for haploid production. However, fertilization and elimination occurred only in Chinese Spring. These observations led to the suggestion that this process is under genetic control.

The crossability of wheat with tetraploid *H. bulbosum* is restricted to a few genotypes (Falk & Kasha, 1981) and is genetically controlled by the genes *Kr*1 and *Kr*2 on chromosome 5B and 5A respectively which are also responsible for determining the crossability with rye, *Secale cereale* (Falk & Kasha, 1983; Sitch *et al.*, 1985; Snape *et al.*, 1979). Wheat genotypes carrying the dominant genes *Kr*1 and *Kr*2 are not crossable with *H. bulbosum*. In interspecific crosses of barley and diploid *H. bulbosum* a dominant gene for partial incompatibility is located on barley chromosome 7 (Pickering, 1983). This chromosome also shows homology with group 5 of wheat chromosomes (Islam & Shepherd, 1982). According to the genealogical pedigrees of wheat cultivars, the crossable wheat genotypes can be traced directly or indirectly to the variety Chinese Spring or to an Asian origin. The data on the crossability with rye further suggest that most of the highly crossable wheat genotypes have their origin in China, Korea and Japan (Riley & Chapman, 1967; Tozu, 1966). In fact, both Japanese and Chinese wheat, in particular local cultivars, have proved to be highly crossable with *H. bulbosum* at a frequency of about 30% seed set (Inagaki & Snape, 1982; Li *et al.*, 1982). The efficiency of wheat haploid production by application of the *bulbosum* method is profoundly influenced by the crossability of *H. bulbosum* onto wheat, so that at present this method is restricted to crossable wheat genotypes.

According to Inagaki (1987), the *bulbosum* technique offers the following advantages compared to anther culture: (1) higher frequency of haploid production; (2) non-occurrence of albinos and negligible number of aneuploids amongst the regenerants; and (3) the production of immature haploid embryos which are interesting material for haploid cell culture.

Laurie & Bennett (1988a,b) carried out wide hybridization with wheat and found fertilization between wheat and maize and/or sorghum and between barley and maize. In the metaphase of zygotes there are 21 wheat chromo-

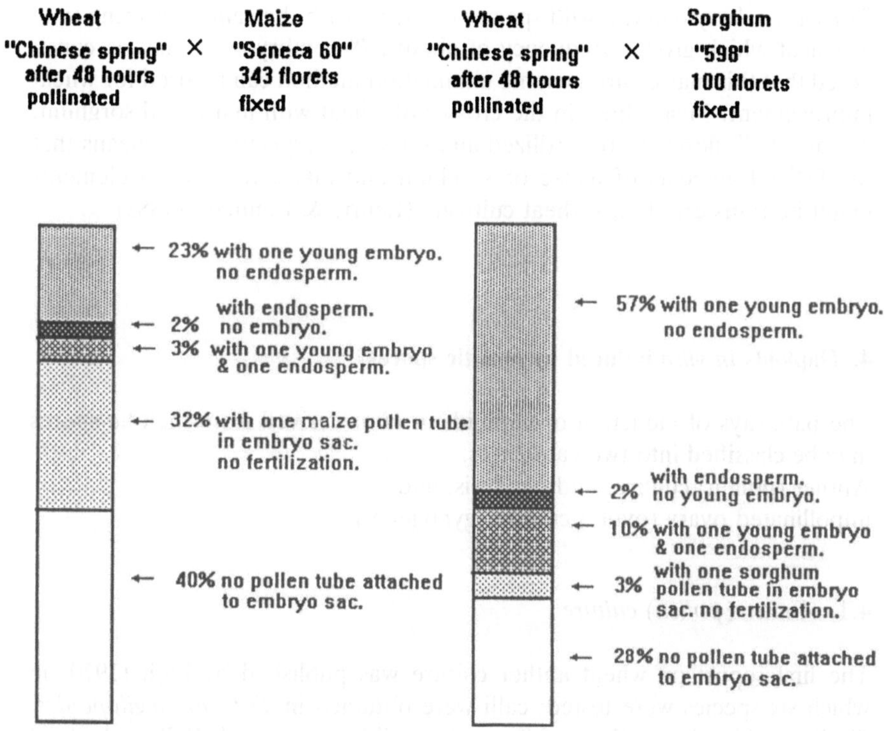


*Figure* 2. Chromosome elimination in wheat crossed with maize and sorghum.

somes or 7 barley chromosomes and 10 maize chromosomes. It indicates that fertilization has been taken place, but the karyotype of the hybrid is unstable. In the initial cell cycle, all maize chromosomes were eliminated. Therefore, wheat haploid plants could be recovered via *in vitro* culture of these wheat haploid embryos. The authors made 26 crosses between wheat and maize, 25 crosses among them were fertilized. The embryological observations were conducted and results are shown in Fig. 2. Similar results of cytoembryological investigations have been obtained by Wang *et al.* (1991).

The maize technique is currently being improved as an alternative method for wheat haploid production (Inagaki & Tahir, 1990; Suenage & Nakajima, 1989; Ushiyama *et al.*, 1991). Meanwhile Kisana *et al.* (1994) have compared two methods, anther culture and wheat × maize method, in a wheat breeding program and found that the wheat × maize technique lead to a higher frequency of haploids and no variation in progenies. But under field conditions, the cross between wheat and maize is difficult due to the difference of their flowering time. Recently, Li (personal communication) found that some wheat cultivars, e.g., Gansu No. 4, proved to be highly crossable with

*Tripsacum dactyloides*, a wild species of maize with the same flowering time as wheat which gives a frequency of about 57% embryos. It may be considered that this maize chromosome elimination method can be used for wheat improvement. Meanwhile, in the crosses of wheat with maize and sorghum, 3% and 10% florets were fertilized and set seed, respectively. It means that the DNA fragments of maize or sorghum and maize transposon elements might be transferred into wheat cultivars (Laurie & Bennett, 1988a).

## 4. Haploids *in vitro* induced by meiotic spores

The pathways of induction of haploids *in vitro* derived from meiotic spores may be classified into two categories:
Anther (pollen) culture: androgenesis, and
unpollinated ovary (ovule) culture: gynogenesis.

### 4.1. *Anther (pollen) culture*

The first report on wheat anther culture was published by Fujii (1970) in which six species were tested; calli were obtained in *Triticum aegilopoides*, *T. dicoccoides*, but anthers of *T. aestivum* did not respond. Pollen plants of wheat were first obtained by Ouyang *et al.* (1973). At that time the regeneration frequency of green plants was only 0.7%.

The success in induction of wheat (*Triticum aestivum*) pollen plants through anther culture was achieved in several laboratories in the early 1970s (Chu *et al.*, 1973; Craig, 1974; Ouyang *et al.*, 1973; Picard & De Buyser, 1973; Research Group 301, 1972). In these years great effort was made to study the methodological and theoretical aspects of wheat anther culture. The pollen plant induction frequency was increased to a level adequate for wheat researchers to apply the anther culture technique in breeding programs.

In recent years, the frequency of microspore embryogenesis in cereals, mainly barley, wheat and rice, has been increased significantly. So now regeneration rates are measured by green pollen plants per anther compared

*Table 1.* Maximum frequencies of green plants reported via anther culture in wheat based on a few selected references

| Reference | Yield of green plants per 100 anthers |
|---|---|
| Liu & Hu (1989) | 117 |
| Chu *et al.* (1990) | 360 |
| Kasha *et al.* (1990) | 322 |
| Orshinsky *et al.* (1990) | 200–455 |

to 100 anthers previously. Table 1 shows the maximum frequencies of green plants obtained via anther culture in wheat.

With such progress, anther/pollen culture has become an extremely efficient system for theoretical and practical applications. Three aspects will be investigated: 1) recovering and creating new types, 2) screening mutants and variants, 3) gene transformation.

There are two pathways of regeneration: organogenesis via a distinct callus phase, or direct embryogenesis (without a callus phase). The embryogenesis route is preferred in plant improvement programs using hybrid material since each regenerant is clearly derived from a single microspore cell. The organogenesis route is generally longer and therefore is more appropriate where mutagenesis is to be applied to the tissue for *in vitro* selection, or where a prolonged culture period is required to maximise the amount of gametoclonal variation induced in the regenerants.

The induction frequency of green pollen plantlets has been steadily increased, and the following factors are considered to be important

1. genotype and physiological state of donor plants,
2. the stage of microspore development,
3. culture media, and
4. culture conditions.

4.1.1. *Genotype and physiological state of donor plants*

The genotype of donor plants has a great influence on the response of anther culture. In earlier studies, significant differences in callus formation using various varieties or crosses were observed.

There is evidence that the response of a genotype is controlled by the genetic constitution of the donor plants. Lazar *et al.* (1987) investigated the anther culture of alien addition lines of cv. Chinese Spring (CS) and found that chromosome 4R promotes the anther response, while 6R and 1R affect the regeneration ability. Szakacs *et al.* (1988) found through analysis of anther culture of CS/Cheyenne substitution line that the induction of callus, regeneration of pollen plants and regeneration of green plants were controlled by multigenes. They are independently heritable characteristics effected by 7A and 1B, 3A and 2D chromosomes, respectively. Agache *et al.* (1988, 1989) also demonstrated that independent multigenes control the above mentioned heritable characteristics during the anther culture procedure. For example there are genotypic differences in relation to pollen callus formation and induction frequency of green pollen plantlets. Cultivars like Xiaoyuan 759 and Kedong 58, characterised by high frequency of regeneration crossed with recalcitrant genotypes lead to $F_1$ hybrids with significantly higher induction rates of pollen callus and green plantlets than the (recalcitrant) parent. This illustrates why these cultivars are also called "bridge varieties".

In addition to the genotype of donor plants, the nature of the wheat cultivars (spring wheat or winter wheat) also affected the regeneration frequency. In general, the induction frequency of pollen callus and green pollen plantlets of spring varieties was much higher than for winter varieties.

Work by Ouyang *et al.* (1989) also proved that conditions under which donor plants were grown could strongly influence the yield of pollen plantlets. The results show that the most suitable culture temperature for anther culture of field-grown material was about 2°C higher than that of greenhouse-grown material. For example, in anther culture of cv. Jinghong 5, the highest yield of pollen plantlets appeared at the culture temperature of 30°C when the anther donor plants were grown in greenhouse compared to 32°C when the donor plants were grown in the field. It was noticed that the anthers of the greenhouse-grown material were more poorly developed than those of the field-grown material. So it is supposed that the badly developed anthers might have weaker metabolism, thus requiring lower culture temperature.

#### 4.1.2. *Pollen developmental stage*

The pollen and anther developmental stage at the time of excision and inoculation is a very important factor for the induction of pollen plants. Some authors pointed out that anthers containing mid- or late-uninucleate microspores were most suitable for anther culture in wheat (Ouyang *et al.*, 1973; Cytological Laboratory of Lanzhou University, 1975; Pan & Kao, 1978). He & Ouyang (1984) have made a precise and systematic study on this subject. It was found that in some genotypes, haploid callus and subsequently haploid green plantlets could be obtained from anthers at various stages from as early as meiosis stage to as late as binucleate stage. But the highest peaks of callus plantlet induction frequency occurred at mid- or late-uninucleate stage in almost all genotypes as reported in previous papers. He and Ouyang (1984) exactly divided the microspore stage into early-uninucleate (including early stage I and early stage II), mid-uninucleate, late-uninucleate and premitosis stage (Fig. 3). These stages are characterized mainly by size, shape and position of the nucleus and by size and presence or absence of the vacuole. The response of anther culture to change of pollen stage is very sensitive, and the yield of pollen callus decreases sharply if anthers at stages other than the mid- or late-uninucleate stage are used.

#### 4.1.3. *Improvement of media components*

The culture media represent a major factor for inducing the development of green plantlets from microspores. Table 2 shows the components of some basic media commonly used for anther culture in cereals. Modifications of carbon and nitrogen sources have significantly improved the response of anther culture.

In early studies, the sucrose level varied between 6–12%; 9% sucrose has been found to be the most appropriate concentration for wheat anther culture to determine not only pollen embryo induction frequency but also regeneration ability (Ouyang, 1986). Further study has shown that 9% was optimal

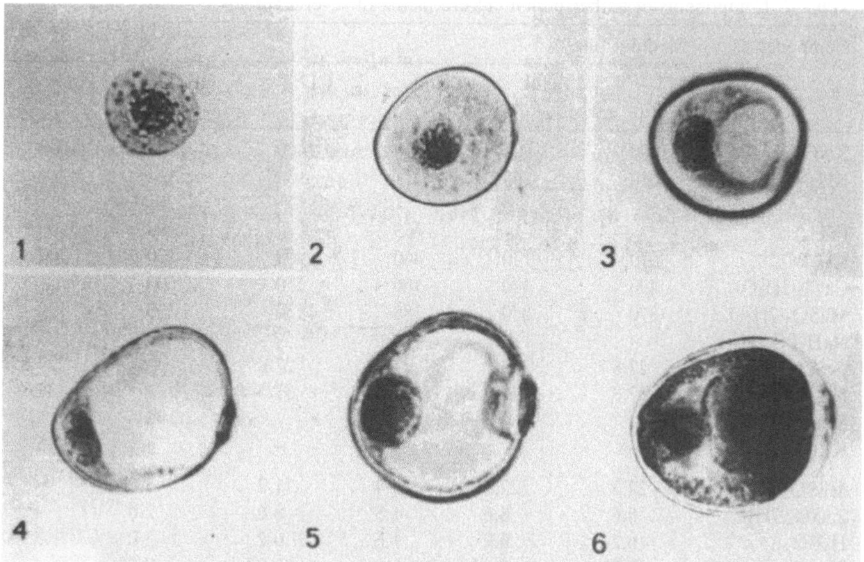


*Figure* 3. Pollen grains of various developmental stages ( × 650): (1) Early uninucleate stage I, without germination pore; (2) early uninucleate stage II; (3) mid-uninucleate stage; (4) late-uninucleate stage; (5) premitosis stage; (6) binucleate stage.

was optimal only for the induction phase, while 2% was sufficient for completion of the embryogenic process (Henry & De Buyser, 1981).

Recently the carbon source in the induction medium was found to have a profound effect on anther culture response. Maltose and other carbohydrates were found to be superior for green plantlet regeneration from barley anthers (Hunter, 1988). Comparing the effects of different carbohydrates on the response to barley anther culture, Hunter carried out experiments using various carbon sources e.g. cellobiose, fructose, glucose, maltose, sucrose, and trehalose to detect the effects of different carbohydrates as a means of improving barley anther culture. He found that high concentration of sucrose and its derivative products, such as glucose and fructose, inhibited the induction of pollen and the formation of young embryos. He also reported that the break-down products of maltose and sucrose differed, which may be a key to the advantage of maltose. It is possible that maltose is degraded more slowly than sucrose, providing a readily metabolizable carbon source over a longer period of culture.

Orshinsky *et al.* (1990) and Last & Brettel (1990) also used maltose instead of sucrose for improving wheat anther culture. The results of their investigation clearly showed that the carbon source used for wheat anther culture can have a significant effect on androgenic response. Maltose was superior to

*Table 2.* Components of anther pollen culture media of several cereals

| Components | Medium (mg/l) | | | | | |
|---|---|---|---|---|---|---|
| | MS | FHG | N6 | C17 | BAC* | Potato II** |
| $KNO_3$ | 1900 | 1900 | 2830 | 1400 | 2600 | 1000 |
| $NH_4NO_3$ | 1650 | 165 | – | 300 | – | – |
| $(NH_4)_2SO_4$ | – | – | 463 | – | 400 | 100 |
| $KH_2PO_4$ | 170 | 170 | 460 | 400 | 170 | 200 |
| $CaCl_2 \cdot 2H_2O$ | 400 | 440 | 166 | 150 | 600 | – |
| $MgSO_4 \cdot 7H_2O$ | 370 | 370 | 185 | 150 | 300 | 125 |
| $NaH_2PO_4 \cdot H_2O$ | – | – | – | – | 150 | – |
| $FeSO_4 \cdot 7H_2O$ | 27.8 | – | 27.8 | 27.8 | – | 27.8 |
| $Na_2EDTA \cdot 2H_2O$ | 37.3 | 40 | 37.3 | 37.3 | – | 37.3 |
| Sequestrene 330Fe | – | – | | – | 40 | – |
| KCl | – | – | – | – | – | 35 |
| $MnSO_4 \cdot 4H_2O$ | 22.3 | 22.3 | 4.4 | 11.2 | 5.0 | – |
| $ZnSO_4 \cdot 7H_2O$ | 8.6 | 8.6 | 1.5 | 8.6 | 2.0 | – |
| $H_3BO_3$ | 6.2 | 6.2 | 1.6 | 6.2 | 5.0 | – |
| KI | 0.83 | 0.83 | 0.8 | 0.83 | 0.8 | – |
| $NaMoO_4 \cdot 2H_2O$ | 0.25 | 0.25 | – | – | 0.25 | |
| $CuSO_4 \cdot 5H_2O$ | 0.025 | 0.025 | – | 0.025 | 0.025 | – |
| $CoCl_2 \cdot 6H_2O$ | 0.025 | 0.025 | – | 0.025 | 0.025 | – |
| Myo-inositol | 100 | 100 | – | – | 2000 | – |
| Thiamine-HCl | 0.4 | 0.4 | 1.0 | 1.0 | 1.0 | 1.0 |
| Pyridoxine-HCl | 0.5 | – | 0.5 | 0.5 | 0.5 | – |
| Nicotinic acid | 0.5 | – | 0.5 | 0.5 | 0.5 | – |
| Glycine | 2.0 | – | 2.0 | 2.0 | – | – |
| Glutamine | – | 730 | – | – | – | 0.5 to 1 g/L |
| Casein hydrolysate | – | – | 500 | 300 | – | – |
| Sucrose | 30,000 | – | 60,000 | 90,000 | 60,000 | 90,000 |
| Glucose | – | – | – | – | 17,500 | – |
| Maltose | – | 62,000 | – | – | – | – |
| Ficoll-400 | – | 200,000 | – | – | 300,000 | – |
| pH | 5.7 | 5.6 | 5.8 | 5.8 | 6.2 | 5.8 |

* BAC3 medium also contains (mol/l) $KHCO_3$ (50), Ascorbic acid (1.0), Citric acid (10), Pyruvic acid (10), $AgNO_3$ (10).

** Potato II medium contains 10% aqueous potato extract.

sucrose both for callus induction and green plantlet regeneration for a range of genotypes. The use of maltose also allowed green plantlets to be recovered from genotypes ($F_1$ 88267, SWP2242) that failed to produce plantlets when anthers were cultured in medium with sucrose. Meanwhile the genotypes used in their study displayed an exceptionally high level of green plantlet regeneration. Cultivar HY368 and CPS hybrids 88281 and 88292 yielded, on average, over 100 green plantlets per 100 anthers cultured. Individual plates of anthers from these genotypes were significantly above these average responses. For example, the best plate of HY368 (containing 20 anthers) produced 91 green plantlets, equivalent to 455 per 100 anthers.

Chu *et al.* (1990) used monosaccharides, especially glucose, instead of sucrose as the carbon source in wheat anther culture. The pollen embryo induction frequency in a liquid medium with 0.21 M glucose was 40–750 embryos per 100 cultured anthers and 2–10 times higher than that in the medium with 0.21 M sucrose depending on the genotypes used. In one wheat genotype ($F_1$, SC8828) more than 360 pollen plants could be produced from 100 cultured anthers. Glucose in the medium greatly enhances pollen embryo and plant production. Compared to glucose, sucrose more or less suppresses wheat pollen embryo development. Sucrose probably induces anther cultures to produce ethylene, which disturbs *in vitro* pollen embryogenesis.

Polysaccharides such as barley starch also improved the response of barley anthers in culture (Sorvari & Schieder, 1987). Starch presumably also serves as a slow release form of glucose, being hydrolysed by amylases produced by the anther wall or by the microspores themselves. According to the abovementioned investigations, salts such as $Na^+$ and $K^+$, amino acids, organic acids, etc., could be more easily absorbed by pollen grains due to the slow metabolism of carbohydrates. These salts and amino acids in pollen cells are important factors for embryogenesis and regeneration of plantlets.

In dedifferentiation medium for anther culture the composition of the nitrogen source is an important factor. The FHG medium, modified by Hunter (1988), significantly increased the frequency of green plantlets in barley. Compared to MS medium, FHG medium differs by replacement of sucrose with maltose; the concentration of $NH_4NO_3$ was also reduced and glutamine was added. Olsen (1988) obtained similar results in MS medium when $NH_4NO_3$ was reduced from 20 to 2 m mol/l and 5.1 m mol/l glutamine was added.

Feng & Ouyang (1989) investigated the effects of nitrogen sources in wheat anther culture. They found that in the dedifferentiation medium $NO_3^-$ affected callus induction and regeneration independent of $NH_4^+$, but the effects of $NO_3^-$ and $NH_4^+$ were additive. They also observed that $KNO_3$ concentration influenced green plant regeneration remarkably: the higher the $KNO_3$ level used, the higher the frequency of green plant regeneration. It was shown that 15–20 m mol/l of $NO_3^-$ was an optimum for callus induction, while 20 m mol/l was most suitable for obtaining high green plant yield. A range of 2–7 m mol/l of $NH_4^+$ concentration was suitable for callus induction, green plant regeneration, and obtaining high green plant yield. A medium without $NO_3^-$ or $NH_4^+$ gave poor regeneration results.

A spectrum of vitamins is routinely included in media, but also often amino acids or an amino acid mixture, such as glycine, glutamine, asparagine, aspartic acid, serine, arginine, alanine, proline, etc. Chu & Hill (1988) utilised an impressive array of minor additives in wheat induction media.

Auxin and the analogues are important growth regulators in plant tissue culture. There have been two trends in using 2,4–D: one is application of low levels or even omission, the other one is adding high concentrations. Both could serve to improve regeneration. According to experience, 2.0 mg/l of 2,4–D is better to be employed together with 0.5 mg/l cytokinin in anther culture of wheat (Liu & Hu, 1989). However, a higher concentration of 2,4–D increased the quality of pollen calli as well as the frequency of calli induction. Subsequently the green plantlet yields were raised significantly in certain genotypes, especially those of poor anther culture response or winter types. High 2,4–D levels showed no harmful effect but there was still a strong influence of the genotype; maybe it will provide a clue for increasing the efficiency of anther culture in poorly responding cultivars of cereals.

For the average number of calli per responding anther (CPRA) in Kedong 58, positive effects of high levels of 2,4–D were achieved by increasing the proportion of responding anthers rather than by altering the CPRA values. In other words, the addition of 2,4–D stimulated many anthers which were originally unresponsive. A possible explanation for this phenomenon is that the receptor of 2,4–D is located in the anther walls and not in the pollen grains. The wall might protect the pollen from the possible harm of high 2,4–D levels and transmit at the same time its stimulation effect to some of the pollen situated near the anther wall, and so lead to initiation. It might be assumed that this receptor is probably the so-called anther wall factor. A similar hypothesis has been proposed for barley.

It is known that autoclaving affects the media composition by degrading some components (Ke-cheng & Bornman, 1991) and causing undesirable reactions between others (Schenk *et al.*, 1991). Filter-sterilised media have been shown to result in higher culture response and regeneration than autoclaved media (Chu & Hill, 1988) so this is now the preferable procedure for media preparation.

#### 4.1.4. *Improvement of culture methods*

Kao (1981) observed that, by adding Ficoll 400 to increase the density of a culture medium, the frequency of plant formation could be considerably increased. Recently Kao and his colleagues (1991) further refined the culture conditions for induction of green plants by anther culture and indicated that cells in callus tissues under anaerobic conditions, as exist in submerged cultures in liquid medium, contained lactate and alcohol dehydrogenase. The accumulation of lactic acid and other organic acids in the tissue may damage the organelles in the cell and result in the formation of albino plants. Thus, for direct embryogenesis from microspores, an ample supply of a fresh, well-buffered medium, and proper aeration for microspore-derived embryos are essential for obtaining high frequencies of green plants. High Ficoll and sucrose levels in liquid medium keep the anthers floating and trap the microspores in the anthers for a sufficiently long time under a condition of sugar starvation. This starvation is essential for stabilization of microspores and eventually leads to androgenesis (Wei *et al.*, 1986).

Another remarkable improvement is the step culture procedure successfully developed in barley that may also have potential for other cereals including wheat. This procedure is briefly described as follows:

1. *Anther starvation*: Anthers dissected from spikes are put into 0.3 M mannitol solution without any medium components and incubated for 3–4 days at 25°C. The duration of starvation is more important than osmotic pressure and mannitol solution free of other carbon sources is essential (Li *et al.*, 1993).
2. *Induction culture*: After 3-day starvation, the mannitol solution is replaced by liquid induction medium. These media have undergone various improvements concerning the nitrogen and carbon source. Lower $NH_4NO_3$ levels have been compensated by raising the glutamine concentration and maltose has replaced sucrose.
3. *Regeneration culture*: After 15–20 days, calli derived from anthers are transferred onto solid regeneration medium containing proline which has been shown to promote the regeneration of green pollen plants in wheat, rice and barley. Temperature is another important factor in this step. When 4 different temperatures (10°C, 15°C, 20°C, 25°C) were compared, the frequency of differentiation and yield of green plantlets varied considerably with the highest at 20°C, followed by 15°C and 10°C, and the lowest at 25°C (Hu *et al.*, 1991).
4. *Robust plantlet culture and transplantation*: MET (pp333, multi-effect triazole, a sort of growth retardant) has played an important role in the continued development of anther-derived plants, because MET at certain concentration is able to induce a strong root system, so the frequency of surviving pollen plants has been greatly increased after transplantation.

4.1.5. *Cultivation of isolated pollen/microspore culture*

Recently, there has been an increased focus on isolated pollen culture (Datta *et al.*, 1990) because pollen culture provides haploid single cells and thus is a promising haplophase tool for basic and applied biotechnological breeding programs. Isolated microspore culture apparently offers a more efficient system to regenerate a random sample from a microspore population than does anther culture. A procedure starting from single microspores passing through embryogenesis and developing at a high frequency into green plants is most advantageous and desirable, because early selection of microspore populations becomes possible. Thus microspores offer the possibility of combining single cell selection procedures with the advantages of a haploid system, i.e., absence of cross-feeding and chimerism, direct expression of all genes independent of their recessive or dominant nature and the possibility, after chromosome doubling, to regenerate homozygous lines directly.

Kasha and his colleagues (1990) described a system for producing large numbers of embryos and calli from mechanically isolated wheat microspores. Four main areas in this wheat microspore culture system were described, namely pretreatment, preculture, isolation procedure and media components.

1. *Pretreatment*: Pretreatment requirements vary with genotypes. The cv. Sinton required a 21 day cold-pretreatment of spikes at 4°C, while the optimum response with cv. Veery S. was achieved when the material was cultured immediately after harvest from the donor plant. Spikes from cv. Chris could be kept in the cold (4°C) for up to seven days.
2. *Preculture*: Preculture of anthers in a macroelement-solution supplemented with mannitol (0.4 M) for 7 days at 28°C has been the most critical factor for inducing a large number of microspores to divide. The majority of microspores isolated after preculture was still at the uninucleate stage.
3. *Isolation procedure*: Anthers were placed in a small amount of mannitol and stirred with a magnetic stirring bar at low speed for approximately 20 min. Most microspores were released into the mannitol and cultured in liquid media.
4. *Medium*: Isolated wheat microspores are very sensitive to 2,4–D and sucrose. Microspores generally plasmolyse and die when these substances are added to the medium. When wheat microspores were cultured in barley shed pollen culture media, FHG, a number initiated divisions. Later, slight modifications to this medium (omitting BA, lowering maltose to 3%, addition of 3% mannitol) greatly increased the number of callus structures formed. Using this system and media, 25% of the cultured microspores in cv. Chris underwent divisions and approximately half of these developed and formed macrostructures (calli and a few embryos). However, in most cases, the structures that developed were translucent and of poor quality. Although system and media still can be improved and refined, it is a promising regeneration technique for isolated microspore culture.

Mejza *et al.* (1993) investigated plant regeneration from isolated microspores of *Triticum aestivum* and developed a new method which can be summarized as follows: wheat microspores from a range of genotypes were isolated without prior anther culture and cultured to regenerate self-fertile plants. The microspores were isolated using a microblender and competent microspores were enriched by gradient centrifugation. Maltose was the sole carbohydrate in the culture medium. Results were improved when spikes were pretreated by immersion of the basal ends of detached heads in water at 25°C for 2 days. This procedure lead to highly reproducible production of plants.

### 4.2. In vitro *culture of unpollinated ovaries (or ovules)*

The culture of unpollinated and unfertilized ovules or ovaries was attempted in various plant materials in the 1950s and 1960s without much success, while an unexpected breakthrough was made in anther culture (Maheshwari, 1950; Maheshwari & Rangeawamy, 1965). San Noeum (1976, 1979) first succeeded in regenerating haploid barley by means of ovary culture. Subsequently, haploid plants or plantlets were raised from cultured ovaries of wheat (Zhu & Wu, 1979), tobacco (Zhu & Wu, 1979; Wu & Chen, 1982), rice (Beauville, 1980; Zhou & Yang, 1980, 1981; Kuo, 1982), barley (Wang & Kuang, 1981; Huang *et al.*, 1982), maize (Ao *et al.*, 1982; Truong-Andre & Demarly, 1984), lily (Gu & Cheng, 1983) and sunflower (Cai & Zhou, 1984). So much success within such a short period indicates that the induction of haploids via *in vitro* culture is not as inaccessible as had earlier been thought. The *in vitro* culture procedure for unpollinated ovaries is similar to that for anther culture as discussed previously. *In vitro* culture of unpollinated ovaries may represent a promising biotechnology for crop improvement. Up to now, it has not been practical owing to low frequency of induction of haploids, and the procedure has to be refined and improved. Considering the number of ovaries is much less than pollen, the potential of anther culture for study and application has much more to be tapped than *in vitro* culture of unpollinated ovaries.

## 5. Application of doubled haploids (DH)

### 5.1. *The possibility of using DH plants in crop improvement*

According to biometrical studies with DH lines Snape (1989) indicated that DH systems have the unique genetic property of allowing completely homozygous lines to be developed from heterozygous parents in a single generation. In self-pollinating species, such as wheat, barley and rice, this property can be used to increase the efficiency of cultivar development. There is also an increase in selection efficiency relative to conventional practices because of an increase in additive genetic variation, an absence of dominance variation and within-family segregation, and a decrease in environmental variation effects through greater replication possibilities.

#### 5.1.1. *Time saving advantage*

Time saving is the most obvious advantage of a DH system because yield and other evaluation trials can be done much sooner than with conventional lines, particularly with winter habit materials, because the requirement for vernalization to initiate flowering limits generation turnover with winter crops.

It also takes less time to build up pure stocks of a new cultivar. Since DH are completely homozygous, all stocks are identical and no purification system is required other than isolation to avoid outcrossing, while in the conventional system stocks are usually derived from a single plant of an advanced

*Table 3.* Expectation of phenotypic variances in different generations derived from a cross between inbred parents (Snape, 1989)

| Generation | Variance |
|---|---|
| $F_2$ (between individual plants) | $V_A + V_D + V_{EI}$ |
| $F_3$ (between family means) | $V_A + 1/2V_D + V_{EP}$ |
| $F_1$ derived DH family means | $2V_A + V_{EP}$ |

$V_A$ = Additive component of variation.
$V_D$ = Dominance component of variation.
$V_{EI}$ = Environmental variance between $F_2$ individuals.
$V_{EP}$ = Environmental variance within $F_3$/dihaploid progenies.

generation, so several generations are required to build up sufficient quantities of seed for release.

5.1.2. *Increasing selection efficiency*

Compared to selection during the early generations of a pedigree program, the homozygosity obtained from using a DH system increases the efficiency of selection both for qualitative, major gene characters, and, in particular, for quantitative characters. Thus, it should be easier to identify superior genotypes and to produce new cultivars when using a DH system.

If the selection of desirable recessive alleles at major gene loci occurs in an $F_2$ population, then only a proportion $(1/4)^n$, where *n* is the number of loci segregating, will have the desirable allelic combination. However, with a DH population, such genotypes will be at a frequency of $(1/2)^n$. Thus, the frequency of fixation in an $F_1$ derived DH population is the square root of the probability in an $F_2$ population.

A remarkable advantage of using DH lines is that greater additive genetic variance is expressed among the recombinants produced from a cross than among the relative $F_2$s and $F_3$s. Another advantage is that dominance variation is absent. These properties are illustrated genetically by comparing the expectations of the variances of $F_1$-derived DHs and the equivalent early conventional generations (Table 3). Thirdly, the environmental variance among $F_2$ individuals ($V_{EI}$) is likely to be greater than that between $F_3$ or DH plots ($V_{EP}$), which are composed of genetically similar individuals. Furthermore, replication can be introduced to reduce $V_{EP}$ so that the breeding values of individual lines can be more accurately assessed.

Conventional early generation plots like $F_3$ or $F_4$ will exhibit genetic differences between individuals within the plots unlike DH plots where all individuals are genetically identical (Table 4). This will make visual selection of desirable lines more difficult in early generations compared to DH plots.

Overall, the different genetic properties of DH populations compared to early conventional generations show that selection efficiency is increased.

Although this DH system allows the greatest time saving compared to a

*Table 4.* Expectation of phenotype variances with plots of conventional and doubled haploid generation (Snape, 1989)

| Generation | Variance |
|---|---|
| Within $F_3$ plots | $1/2V_A + 1/2V_D + V_{EI}$ |
| Within $F_4$ plots | $1/4V_A + 1/4V_D + V_{EI}$ |
| Within doubled haploid plots | $V_{EI}$ |

conventional pedigree program, it also has two disadvantages. One is that a completely random sample of gametes is fixed and, thus, genetic drift will result in a high rejection frequency since only a small proportion of lines will fulfill desired criteria. For example, if only a single gene, e.g., for disease resistance, is segregating between the parents, then half of the population is expected to carry the undesirable, susceptible allele. If *n* major gene loci are segregating, only a proportion of $(1/2)^n$ of the DHs will contain the desirable combinations of alleles. Clearly, even for small number of loci this proportion will be small.

The second disadvantage of DH extraction from an $F_1$ hybrid is that only one cycle of recombination is allowed between the parental genomes before fixation. Thus, the DH population will be in linkage disequilibrium if linkages between genes are important components of variation.

The remedy for these disadvantages is to delay haploidization until the $F_2$ or $F_3$ generation and to practice selection prior to haploid production. For example, $F_2$ individuals can be classified for major gene characters and also selected visually for quantitative characters of high heritability, such as flowering time and plant height. Only desirable plants are then used for DH production. Such a scheme is illustrated in Fig. 4. This scheme does not allow much saving in time compared to an $F_1$ system, but does give increased selection efficiency for both major genes and quantitative characters. This should ensure a high frequency of lines with desired levels of performance and, consequently, large population size should not be necessary for genetic advance. Since more cycles of recombination are allowed, there will also be less linkage disequilibrium than in an $F_1$-derived population. More advanced generations can be used in this way, thereby giving more opportunity for selection prior to haploidization.

In practice, to overcome these disadvantages of the DH system, Zhang *et al.* (1983) developed an assembled breeding method which is composed of one cycle of anther culture followed by sexual hybridization between different genotypes of pollen plants and another cycle of anther culture of the selected sexual hybrids. By using this method, some polygenic traits have been successfully improved. An example is the release of the cold-resistant rice cultivar named Hua Han Zao. In addition, Hu *et al.* (1985) developed a method combining anther culture with composite crossing, where the linkage

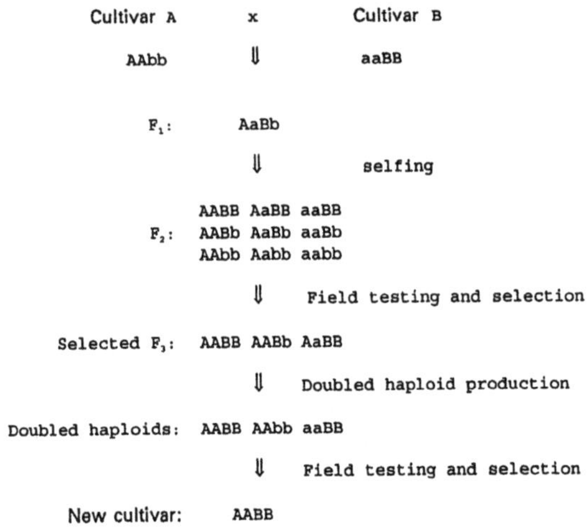


*Figure* 4. The selected $F_3$ doubled haploid system for varietal production in self-pollinating crops (Snape, 1989).

between genes might be broken. By this method, a winter wheat cultivar with desirable characters, Jinghua No.1, has been released.

Based on the above discussion the advantages of DH lines can be exploited very successfully with self-pollinating crops. In such programs different filial generations can be chosen as the parental material for haploidization, although the most common system is to use $F_1$ hybrids thereby fixing recombination between the two parental genomes at the earliest opportunity. As every DH produced is a potential cultivar, field selection is practiced to identify those lines with the desirable combination of characters.

### 5.2. *Developing and releasing new cultivars of wheat*

In 1973 Ouyang *et al.* first obtained DHs of wheat (*Triticum aestivum*) by means of anther culture. Since that time a set of new cultivars (strains) has been developed and released by scientists all over the world, e.g., the Chinese wheat cv. Jinghua No. 1 and the French wheat cv. Florin.

Up to 1991 in China 21 new cultivars (strains) have been developed and released into production using anther culture. They have good agronomic characteristics with high yield and wide adaptation. The amount of acreage under cultivation of these cultivars has reached more than 1,000,000 ha. Table 5 shows the cultivars approved by the Variety Evaluation Committee with acreage more than 27,000 ha. Current wheat breeding laboratories combine anther culture with conventional breeding programs as a routine method for wheat improvement in China.

*Table 5.* Wheat cultivars released in production via anther culture (Hu, 1991)

| Name of cultivar | Breeding and research institute |
|---|---|
| Jinghua No. 1 | Hu, D.F., Laboratory of Plant Cell Engineering, Beijing Academy of Agricultural Sciences |
| Jinghua No. 3 | ibid. |
| Jinghua No. 5 | ibid. |
| Huapei 764 | Academy of Genesu Agricultural Sciences |
| Zheng Chun No. 11 | Institute of Agriculture of Gansu Zheng Ye Region |
| Gan Chun No. 16 | Gansu Agricultural University |
| Yu Mai No. 6 | Luo Yang Agricultural Institute |
| Kui Hua No. 2 | Sin Jing Nong Qi Shi Agricultural Institute |
| Xia Mai No. 1 | Laboratory of Genetics, Institute of Wheat, Henan Academy of Agriculture |
| Anther Culture 28 | Zhao, Y. L. *et al.*, Laboratory of Genetics, Institute of Wheat, Henan Academy of Agriculture |
| Hua 555 | Wang P. *et al.*, Hebei Academy of Agriculture |

### 5.3. *Chromosome engineering in Triticeae using pollen-derived plants (CETPP)*

This system has been established by Han Hu and his group. It combines chromosome engineering with anther culture and modified identification methods transfering the desirable chromosomes (genes) into cultivars and thus to create new strains of wheat. As chromosome substitution, addition and translocation lines have been obtained, this CETPP method has significant potential as a tool in germplasm enhancement (Hu, 1992).

### 5.4. *Genetic manipulation*

As microspore culture is a single cell system, it makes selection on the single cell level possible and furthermore offers new prospects for genetic manipulation, e.g., mutagenesis and transformation. Direct gene transfer by microinjection into isolated multicellular pollen embryos offers the possibility of transgenic plant formation of all cereals including wheat by using culture of isolated pollen having high regeneration efficiency (Potrykus *et al.*, 1985; Potrykus, 1988). It can be expected that new and interesting information concerning genetic manipulation using microspore culture of wheat will be obtained in the near future.

### 5.5. *Use of DH populations*

A new field of haploid usage is molecular genome identification, particularly for QTL analysis. DH populations are an important tool to obtain reproducible DNA polymorphisms in barley (Heun *et al.*, 1991) and rice

(McMouch *et al.*, 1991; Xu *et al.*, 1994), etc. One method for generating an RFLP map of these cereals is based on DH populations.

## 6. Conclusion

There are three pathways to induce haploid sporophytes of higher plants:

I. From meiotic spores by anther (pollen) culture and unpollinated ovary (ovule) culture;

II. From reduced zygotes by chromosome elimination; and

III. From male and female gametes by isolation and manipulation of reproductive cells and protoplasts.

Among them, approaches I and II have been developed and utilised in production, with great progress. This progress has paved the way for fundamental studies and practical utilization of DHs.

The frequency of microspore embryogenesis in cereals has been increased significantly. Up to now in several genotypes of barley and wheat one anther could produce more than one green plantlet. The success is calculated by green pollen plants per anther in comparison with the previous rate of per 100 anthers. But one anther of cereals contains more than 2000–3000 pollen grains, so there is a much greater potential for anther-pollen (microspore) culture.

The initial period of culture might be a starvation period in which the process of dedifferentiation and the acquisition of embryogenic capacity in cultured pollen occur (Heberle-Bors, 1989; Li *et al.*, 1993). The direction of pollen development can be influenced during this period, by switching its course away from gametophytic development to sporophytic development (Kyo and Harada, 1989). So anther-pollen culture can be a model system for studying cell differentiation and other fundamental biological problems.

In practice, DHs combined with different breeding methods, such as conventional breeding, chromosome engineering, mutagenesis and wide hybridization will efficiently recover and screen a lot of recombinants and variants. Meanwhile it can also create new types which are difficult to obtain by conventional methods for wheat improvement.

In addition, microspore culture in combination with genetic transformation may result in an increased research effort in this field.

## 7. Acknowledgements

The author is very grateful to B. Winkelmann for critical reading of the manuscript, J.K. Jing for photographic work and Y. Li for typing.

## 8. References

Agache, S., J. de Buyser, Y. Henry & J.W. Snape (1988) Studies on the genetic relationship between anther culture and somatic tissue culture ability in wheat. Plant Breed. 100: 26–33.

Agache, S., B. Bachelier, J. de Buyser, Y. Henry & J.W. Snape (1989) Genetic analysis of anther culture response in wheat using aneuploid, chromosome substitution and translocation lines. Theor. Appl. Genet. 97:7–11.

Ao, G.M., S.X. Zhao & G.H. Li (1982) *In vitro* induction of haploid plantlets from unpollinated ovaries of corn (*Zea mays* L.) Acta Genet. Sin. 9: 281–283 (in Chinese with English abstract).

Barclay, I.R. (1975) High frequencies of haploid production in wheat (*Triticum aestivum*) by chromosome elimination. Nature 256: 410–411.

Beauville, A.M. (1980) Obtention d'haploïdes *in vitro* à partir d'ovaires non fecondés de riz, *Oryza sativa* L. C.R. Acad. Sci., Paris 290: 489–492.

Blakeslee, A.F., J. Belling, M.E. Farnham & A.D. Bergner (1922) A haploid mutant in the jimson weed *Datura stramonium*. Science 55: No. 1433.

Cai, D.T. & C. Zhou (1984) *In vitro* induction of haploid embryoids and plantlets from unpollinated young florets and ovules of *Helianthus annuus* L. Kexue Tongbao 29(5): 680–682.

Chen, Y. & L. Li (1978) Investigation and utilization of pollen derived haploid plants in rice and wheat. In: Proc. of Sino-Australian Symposium on Plant Tissue Culture, pp. 199–212. Science Press, Beijing.

Chu, C.C., C.C. Wang, C.S. Sun, N.F. Chien, K.C. Yin & C. Hsu (1973) Investigations on the induction and morphogenesis of wheat (*Triticum aestivum*) pollen plants. Acta Bot. Sin. 15: 1–11 (in Chinese with English abstract).

Chu, C.C. & R.D. Hill (1988) An improved anther culture method for obtaining higher frequency of pollen embryoids in *Triticum aestivum* L. Plant Sci. 55: 175–181.

Chu, C.C., R.D. Hill & A.L. Brule-Babel (1990) High frequency of pollen embryoid formation and plant regeneration in *Triticum aestivum* L. on monosaccharide containing media. Plant Sci. 66: 255–262.

Craig, I.L. (1974) Haploid plants (2n = 21) from *in vitro* anther culture of *Triticum aestivum*. Can. J. Genet. Cytol. 16: 697–700.

Cytological Laboratory, Dept. Biol. Lanchows Univ. Div. Bot., Agric. Univ. Kansu (1975) Studies on the conditions for induction of wheat (*Triticum aestivum*) pollen plants. Acta Genet. Sin. 4: 302–310 (in Chinese with English abstract).

Datta, S.K., I. Potrykus, M. Bolik & G. Wenzel (1990) Culture of isolated pollen of wheat (*Triticum aestivum* L.). In: Y.P.S. Bajaj (Ed.), Biotechnology in Agriculture and Forestry, Vol. 13. Wheat, pp. 435–447. Springer-Verlag, Berlin.

Falk, D.E. & K.J. Kasha (1981) Comparison of the crossability of rye (*Secale cereale*) and *Hordeum bulbosum* onto wheat (*Triticum aestivum*). Can. J. Gent. Cytol. 23: 81–88.

Falk, D.E. & K.J. Kasha (1983) Genetic studies of the crossability of hexaploid wheat with rye and *Hordeum bulbosum*. Theor. Appl. Genet. 64: 303–307.

Feng, G.H. & J.W. Ouyang (1989) Studies on effects of different nitrogen sources in anther culture medium of wheat. In: C.S. Kuo *et al.* (Eds.), Recent Advances on Studies of Applied and Fundamental Aspects of Plant Cell Engineering, pp. 126–132. Science J. Press, Beijing.

Fujii, T. (1970) Callus formation in wheat anther. Wheat Inf. Serv. 31: 1–2.

Gu, M.G. (1986) Cytogenetic stability and variability of calli and cell clones originated from maize pollen and their regenerated plants. In: H. Hu & H.Y. Yang (Eds.), Haploids of Higher Plants *in Vitro*, pp. 79–90. China Academic Publishers, Beijing/Springer Verlag, Berlin.

Gu, Z.P. & K.C. Cheng (1983) *In vitro* induction of haploid plantlets from unpollinated ovaries of lily and its embryological observations. Acta Bot. Sin. 25: 24–28 (in Chinese with English abstracts).

He, D.G. & J.W. Ouyang (1984) Callus and plantlet formation from cultured wheat anthers at different developmental stages. Plant Sci. Lett. 33: 71–79.

Heberle-Bors, E. (1989) Isolation pollen culture in tobacco: Plant reproductive development in a nutshell. Sex. Plant Reprod. 2: 1–10.
Henry, Y. & J. de Buyser (1981) Float culture of wheat anthers. Theor. Appl. Genet. 60: 77–79.
Heun, M., A.E. Kennedy, J.A. Anderson, N.L.V. Lapitan, M.E. Sorrelis & S.D. Tanksley (1991) Construction of a restriction fragment length polymorphism map for barley. Genome 34: 437–447.
Hu, D.F., Z.D. Yuan, Y. Tang & J.P. Liu (1985) Jinghua No 1., a winter wheat variety derived from pollen sporophyte. Sci. Sin. 28: 733–745.
Hu, D.F. (1991) Application of anther culture technique in crop improvement. In: The Chinese Association for Sciences (Ed.), The Investigation of High and New Technologies in Agricultural Application, pp. 17–23. China Academic Publishers, Beijing.
Hu, H., T.Y. Hsi & S.G. Chia (1978) Chromosome variation of somatic cells of pollen calli and plants in wheat (*Triticum aestivum* L.). Acta Genet. Sin. 5: 23–29 (in Chinese with English abstracts).
Hu, H., T.Y. Hsi & J.W. Ouyang (1980) Chromosome variation of pollen mother cell of pollen-derived plants in wheat (*Triticum aestivum* L.). Sci. Sin. 23: 905–912.
Hu, H., W.Z. Li & J.K. Jing (1991) Anther/pollen *in vitro* culture of barley. Barley Genet. Newsl. 6: 203–205.
Hu, H. (1992) Germplasm enhancement by anther culture in Triticeae. Hereditas 116: 151–154.
Huang, Q.F., H.Y. Yang & C. Zhou (1982) Embryological observations on ovary culture of unpollinated young flowers in *Hordeum vulgare* L. Acta Bot. Sin. 24: 295–300 (in Chinese with English abstract).
Hunter, C.P. (1988) Plant regeneration from microspores of barley *Hordeum vulgare* L. Ph.D. Thesis. Wye College, University of London. Springer Verlag, Berlin.
Inagaki, M.N. (1987) Variation in plant height of doubled haploid lines of wheat derived from intergeneric cross with *Hordeum bulbosum* L. Jpn. J. Breed. 37: 275–282.
Inagaki, M.N. & J.W. Snape (1982) Frequencies of haploid production in Japanese wheat varieties crossed with tetraploid *Hordeum bulbosum* L. Jpn. J. Breed. 32: 341–347.
Inagaki, M.N. & M. Tahir (1990) Comparison of haploid production frequencies in wheat varieties crossed with *Hordeum bulbosum* L. and maize. Jpn. J. Breed. 40: 209–216.
Islam, A.K.M.R. & K.W. Shepherd (1982) Wheat-barley addition lines: their use in genetic and evolutionary studies of barley. Barley Genet. Newsl. 4: 729–739.
Kao, K.N. (1981) Plant formation from barley anther cultures with Ficoll media. Z. Pflanzenphysiol. 103: 437–443.
Kao, K.N., M. Saleem, S. Abrams, M. Pedras, D. Horn & C. Mallard (1991) Culture conditions for induction of green plants from barley microspores by anther culture methods. Plant Cell. Rep. 9: 595–601.
Kasha, K.J. & K.N. Kao (1970) High frequency of haploid production in barley (*Hordeum vulgare* L.). Nature 225: 874–875.
Kasha, K.J., A. Ziauddin & U.H. Cho (1990) Haploids in cereal improvement: Anther and microspore culture. In: J.P. Gustafson (Ed.), Gene Manipulation in Plant Improvement II, Stadler Genet. Symp. pp. 213–235. Plenum Press, New York.
Ke-cheng, H. & C.H. Bornman (1991) An oxygen electrode assay to determine autoclaved-induced toxicity in tissue culture media. Physiol. Plant. 81: 55–58.
Kimber, G. & R. Riley (1963) Haploid angiosperms. Bot. Rev. 29: 480–531.
Kisana, N.S., K.K. Nkongolo, J.S. Quick & D.L. Johnson (1993) Production of doubled haploids by anther culture and wheat × maize method in a wheat breeding programme. Plant Breed. 110: 96–102.
Kuo, C.S. (1982) The preliminary studies on culture of unfertilized ovaries of rice *in vitro*. Acta Bot. Sin. 24: 33–38 (in Chinese with English abstract).
Kyo, M. & H. Harada (1986) Control of the developmental pathway of tobacco pollen *in vitro*. Planta 168: 427–432.

Last, D.I. & R.I.S. Brettell (1990) Embryo yield in wheat anther culture is influenced by choice of sugar in the culture medium. Plant Cell. Rep. 9: 14–16.

Laurie, D.A. & M.D. Bennett (1988a) The production of haploid wheat plants from wheat × maize crosses. Theor. Appl. Genet. 76: 393–397.

Laurie, D.A. & M.D. Bennett (1988b) Cytological evidence for fertilization in hexaploid wheat × sorghum crosses. Plant Breed. 100: 73–82.

Lazar, M.D., T.H.H. Chen, G.T. Scoles & K.K. Kartha (1987) Immature embryo and anther culture of chromosome addition lines of rye in Chinese spring wheat. Plant Sci. 51: 77–81.

Li, D.W., Z.Y. He & Q.D. Hu (1982) The crossability of *Triticum aestivum* with tetraploid *Hordeum bulbosum*. Ann. Rep. Inst. Genet., 1981. Acad. Sin.: 136–138.

Li, W.Z., J.K. Jing, G.H. Yao & H. Hu (1993) The effects of genotypes and mannitol pretreatment of high frequency androgenesis in barley. Chinese Sci. Bull. 38 (2): 151–155.

Liu, C.H. & H. Hu (1989) High frequency of androgenesis in wheat (*Triticum aestivum*). Genet. Manip. Plants 5 (2): 24–28.

Maheshwari, P. (1950) An Introduction to the Embryology of Angiosperms. McGraw-Hill, New York.

Maheshwari, P., N.S. Rangeawamy (1965) Embrology in relation to physiology and genetics. In: R.D. Preston (Ed.), Advances in Botanical Research Vol. 2, pp. 219–321. Academic Press, London.

Maheshwari, S.C., A. Raishid & A.K. Tyagi (1983) Anther pollen culture for production of haploids and their utility. IAPTC Newsl. 41: 2–7.

McMouch, S.R. & S.D. Tanksley (1991) Development and use of RFLP in rice breeding and genetics. In: G.S. Khush and G.T. Toenniessen (eds.) Rice Biotechnol. pp. 109–133. C.A.B. International, UK.

Mejza, S.J., V. Morgant, D.E. Dibona & J.R. Wong (1993) Plant regeneration from isolated microspores of *Triticum aestivum*. Plant Cell. Rep. 12: 149–153.

Olsen, F.L. (1988) Induction of microspore embryogenesis in cultured anthers of *Hordeum vulgare*. The effects of ammonium nitrate glutamine and asparagine as nitrogen sources. Carlsberg Res. Commun. 52: 393–404.

Orshinsky, B.R., L.J. McGregor, G.I.E. Johnson, P. Hucl & K.K. Kartha (1990) Improved embryoid induction and green shoot regeneration from wheat anthers cultured in medium with maltose. Plant Cell. Rep. 9: 365–369.

Ouyang, T.W., H. Hu, C.C. Chuang & C.C. Tseng (1973) Induction of pollen plants from anther of *Triticum aestivum* L. cultured *in vitro*. Sci. Sin. 16(1): 79–95.

Ouyang, J.W. (1986) Induction of pollen plants in *Triticum aestivum*. In: H. Hu & H. Yan (Eds.), Haploids of Higher Plants *in Vitro*, pp. 26–41. Springer-Verlag, Berlin.

Ouyang, J.W, D.G. He, G.H. Feng & S.E. Jia (1989) The response of anther culture to culture temperature varies with growth conditions of anther-donor plants. In: C.S. Kuo *et al.* (Eds.), Recent Advance on Studies of Applied and Fundamental Aspects of Plant Cell Engineering, pp. 122–125. Scientific J. Press, Beijing.

Pan, C.L. & K.H. Kao (1978) The induction of wheat pollen embryo and the influence of some factors on its frequency of induction. In: Proc. Symp. Plant Tissue Culture, pp. 133–142. Science Press, Beijing.

Picard, E. & J. de Buyser (1973) Obtention de plantlets haploïdes de *Triticum aestivum* L. à partir de cultures d'anthères *in vitro*. C.R. Acad. Sci. Paris 277: 1463–1466.

Pickering, R.A. (1983) The location of a gene for incompatibility between *Hordeum vulgare* L. and *Hordeum bulbosum* L. Heredity 51: 455–459.

Potrykus, I., J. Paszkowski, M. Saul, S. Kruger-Lebus, T. Muller, R. Schocher, I. Negrutiu, P. Kunzler & R.D. Shillito (1985) Direct gene transfer to protoplasts: an efficient and generally applicable method for stable alterations of plant genomes. In: M. Freeling (Ed.), Plant Genetics, pp. 181–199. A.R. Liss, New York.

Potrykus, I. (1988) Direct gene transfer to plants. In: Application of Plant Cell Tissue Culture (Ciba Foundation Symposium 137), pp. 144–162. John Wiley Sons, New York.

Research Group 303, Institute of Genetics, Academia Sinica (1972) A preliminary report on

the induction of pollen plants from rice and wheat. Genet. Commun. (Beijing) 1: 1–4 (in Chinese).
Riley, R. & V. Chapman (1967) The inheritance in wheat of crossability with rye. Genet. Res. 9: 259–267.
San Noeum, L.H. (1976) Haploïdes d'*Hordeum vulgare* L. par culture *in vitro* d'ovaires non fecondés. Ann. Amelior. Plant. 26: 751–754.
San Noeum, L.H. (1979) *In vitro* induction of gynogenesis in higher plants. In: Proc. Conf. Broadening Genet. Base Crops, pp. 327–329. Pudoc, Wageningen.
Schenk, N., K.C. Hsiao & C.H. Bornman (1991) Avoidance of precipitation and carbohydrate breakdown in autoclaved plant tissue culture media. Plant Cell. Rep. 10: 115–119.
Sitch, L.A., J.W. Snape & S.J. Firman (1985) Intrachromosomal mapping of crossability genes in wheat (*Triticum aestivum*). Theor. Appl. Genet. 70: 309–314.
Snape, J.W., V. Chapman, J. Moss, C.E. Blanchard & T.E. Miller (1979) The crossabilities of wheat varieties with *Hordeum bulbosum*. Z. Pflanzenzüchtg. 85: 200–204.
Snape, J.W. (1989) Doubled haploid breeding: theoretical basis and practical applications. In: A. Mujeeb-Kazi & L.A. Sitch (Eds.), Review of Advances in Plant Biotechnology, 1985–1989, pp. 19–30. International Maize and Wheat Improvement Center. International Rice Research Institute, Manila.
Sorvari, S. & O. Schieder (1987) Influence of sucrose and melibiose on barley anther cultures in starch media. Plant Breed. 99: 164.
Suenaga, K. & K. Nakajima (1989) Efficient production of haploid wheat (*Triticum aestivum*) through crosses between Japanese wheat and maize (*Zea mays*). Plant Cell Rep. 8: 263–266.
Szakacs, E., K. Geza, P. Jancs & B. Beata (1988) Substitution analysis of callus induction and plant regeneration from anther culture in wheat (*Triticum aestivum* L.). Plant Cell. Rep. 7: 127–129.
Szarejko, I. & K.J. Kasha (1989) Induction of anther culture derived doubled haploids in barley. Proc. FAO/IAEA Research Coord. Meeting, Katowice, Poland, July 1988 (in press).
Tozu, T. (1966) Crossability between wheat and rye. Seiken Ziho 18: 33–38.
Truong-Andre, I. & Y. Demarly (1984) Obtaining plants by *in vitro* culture of unfertilized maize ovaries (*Zea mays* L.) and preliminary studies on the progeny of a gynogenetic plant. Z. Pflanzenzüchtg. 92: 309–320.
Ushiyama, T., T. Shionizu & T. Kuwabara (1991) High frequency of haploid production of wheat through intergeneric cross with teosinte. Jpn J. Breed. 41: 353–357.
Wang, C.C. & B.J. Kuang (1981) Induction of haploid plants from the female gametophyte of *Hordeum vulgare* L. Acta Bot. Sin. 23: 329–330 (in Chinese with English abstract).
Wang, J.L., C.S. Sun, T.G. Lu, A. Fang, H.R. Cui, S.Z. Cheng & C. Xiang (1991) Fertilization and embryological development in wheat × maize crosses. Acta Bot. Sin. 33(9): 674–679.
Wei, Z.M., M. Kyo & H. Harada (1986) Callus formation and plant regeneration through direct culture of isolated pollen of *Hordeum vulgare* cv. Sabarlis. Theor. Appl. Genet. 72: 252–255.
Wu, B.J. & K.C. Chen (1982) Cytological and embryological studies on haploid plant production from cultured unpollinated ovaries of *Nicotiana tabacum* L. Acta Bot. Sin. 24: 125–129 (in Chinese with English abstract).
Xu, J.C., L.H. Zhu, Y. Chen, C.F. Lu & H.W. Cai (1994) Construction of a rice molecular linkage map using a doubled haploid population. Acta Genet. Sin. 21: 205–214.
Yang, H.Y. & C. Zhou (1992) Experimental plant reproductive biology and reproductive cell manipulation in higher plants: Now and the future. Amer. J. Bot. 79: 354–363.
Zenkteler, M. & J. Straub (1979) Cytoembryological studies on the process of fertilization and the development of haploid embryos of *Triticum aestivum*. (2n = 42) after crossing with *Hordeum bulbosum* (2n = 14). Z. Pflanzenzüchtg. 82: 36–44.
Zhang, Z.H., Z.L. Zheng & W.G. Sun (1983) Breeding of new rice varieties through multiple recombination between lines of pollen plants. In: J.H. Shen *et al.* (Eds.), Studies on Anther Culture Breeding in Rice, pp. 1–6. Agric. Press, Beijing (in Chinese with English abstract).
Zhou, C. & H.Y. Yang (1980) *In vitro* induction of haploid plantlets from unpollinated young ovaries of *Oryza sativa* L. Acta Genet. Sin. 7: 287–288 (in Chinese).

Zhou, C. & H.Y. Yang (1981) Induction of haploid rice plantlets by ovary culture. Plant Sci. Lett. 20: 231–237.

Zhu, Z.C. & H.S. Wu (1979) *In vitro* induction of haploid plantlets from the unpollinated ovaries of *Triticum aestivum* and *Nicotiana tabacum* L. Acta Genet. Sin. 23: 499–501 (in Chinese with English abstract).

# 4. Haploidy in barley

B.P. FORSTER and W. POWELL

**Contents**



## 1. Introduction

Barley was one of the first crops to be cultivated by man (Zohary & Hopf, 1988), and today ranks as the fifth most productive crop in the world. Factors which have contributed to its successful cultivation include its tolerance to abiotic stresses; high and low temperatures, drought, mineral deficiencies and toxicities, and its relatively short life cycle. These factors enable barley to be grown in extreme environments not suited to other crops. The main harvestable product is the grain which is used for animal feed, brewing, malting, and as a human food. The straw and foliage are also important sources of animal fodder.

Cultivated barley, *Hordeum vulgare* L., is the only cultivated species in the genus *Hordeum* which encompasses some 30 species (von Bothmer, 1992). The genus forms part of the sub-tribe, the Triticeae, which contains other important small grain cereals; bread wheat, macaroni wheat, rye and triticale. The Triticeae in turn forms part of the grass family, the Poaceae.

Barley as a crop was developed from the domestication of the wild species *H. spontaneum* C. Koch some 10,000 years ago in the Fertile Crescent (Zohary & Hopf, 1988). Fertile hybrids are easily produced between *Hordeum vulgare* and its progenitor, *H. spontaneum*, and there are no barriers to recombination. This has lead some taxonomists to group *H. vulgare* and *H. spontaneum* into one species. In addition, both are diploid and have the same genome symbol ($2n = 2x = 14$, HH). The two species form the primary gene pool for genetic improvement of the crop. *Hordeum bulbosum*, which

*S.M. Jain, S.K. Sopory & R.E. Veilleux (eds.),* In Vitro *Haploid Production in Higher Plants, Vol. 4,* 99–115.

occasionally forms hybrids and shares a common genome (HH) with *H. vulgare*, forms a secondary gene pool. The other *Hordeum* species form a tertiary gene pool.

Cultivated barley is an inbreeder and crop improvement programmes aim to produce superior lines which are homozygous and as a consequence, true breeding. This is traditionally achieved by crossing two or more complementary genotypes of *H. vulgare* followed by repeated selfing and selection cycles for desired traits (Rasmusson, 1992). Occasionally, plant breeders have used *H. spontaneum* as a source of novel genes for barley improvement, e.g., powdery mildew resistance (Thomas *et al.*, 1987). In such cases, repeated backcrossing to the crop species is necessary to eliminate weedy characteristics inherited from the wild species. Wider crosses have not been exploited in this way since wider hybridisation is more difficult. The hybrids produced are often sterile or non-viable, or the cross results in the production of haploids (Fedak, 1992). This latter phenomenon was observed to occur at high frequency in crosses between *H. vulgare* and *H. bulbosum* by Kasha & Kao (1970) who quickly realised the potential of the technique (see later sections).

## 2. Methods of producing haploids in barley

Haploid plants are defined as sporophytes which carry the gametic complement of chromosomes (Kasha & Reinbergs, 1981). Haploids of barley *H. vulgare* are monoploid, and therefore carry seven single chromosomes ($n = x = 7$, H; Fig. 1a) as opposed to the seven pairs of chromosomes found in the diploid ($2n = 2x = 14$, HH; Fig. 1b). The physiology/morphology of barley haploids is similar to normal barley plants except that they are shorter in stature and sterile (Fig. 2).

There are generally two methods of producing haploids in barley. The first uses gametic cells as the source of haploid material. Here *in vitro* culture is used to divert the normal gametophytic development of eggs and pollen grains into haploid plants. This can be done by culturing tissues such as ovaries or anthers which contain developing gametophytes or alternatively the gametic cells can be isolated and cultured directly. The second technique exploits the *in vivo* production of haploid embryos. This can occur in barley plants carrying a haploid inducer gene, *hap*, or as a result of crossing with another species.

### 2.1. *Ovary culture*

San Noeum (1976) was the first to describe the development of haploid plants from unfertilised eggs in ovary cultures of barley. Despite improvements in culturing techniques (Yean-San, 1987) the production of haploid plants by this method has remained very inefficient (0.2–1.4% of cultured ovules

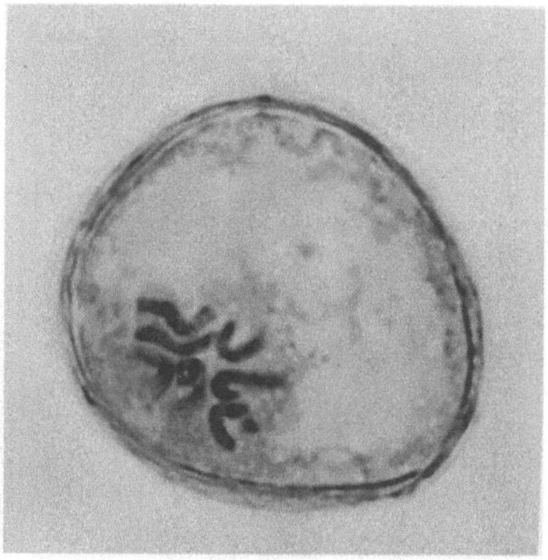

a

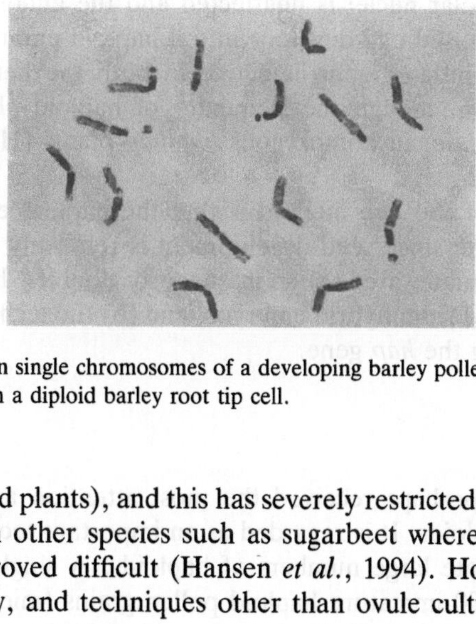

b

*Figure 1.* (a) Seven single chromosomes of a developing barley pollen grain. (b) Seven pairs of chromosomes from a diploid barley root tip cell.

produce haploid plants), and this has severely restricted its use. Ovule culture is important in other species such as sugarbeet where anther culture techniques have proved difficult (Hansen *et al.*, 1994). However, this does not apply to barley, and techniques other than ovule culture have been developed.

### 2.2. *Haploid initiator gene*

The mutant haploid initiator gene, *hap*, was described by Hagberg & Hagberg (1980). The function of the *hap* gene is to prevent fertilisation of the egg.

*Figure 2.* Comparison of the morphology of haploid and diploid barley plants. Haploid plants are shorter and sterile (sterility encourages regrowth of tillers).

Fertilisation of the polar nuclei is unaffected and the endosperm develops normally. The unfertilised egg develops into a haploid embryo (Mogensen, 1982). The haploid initiator gene is active in both the heterozygous and homozygous condition, though the frequency of haploid plant production can be four times greater in homozygous *hap/hap* plants (Hagberg & Hagberg, 1987).

A big advantage of the *hap* method is that the haploid embryos do not require *in vitro* culture since seed development is relatively normal. Disadvantages of the technique are: (a) spontaneously doubled haploids cannot be easily distinguished from hybrid embryos, and (b) the technique is limited to genotypes carrying the *hap* gene.

### 2.3. *Anther culture*

Anther culture has perhaps received the most attention as a method for producing barley haploids. It is regarded as an important technique as there is potential to generate large numbers of haploids. A single barley anther contains roughly 3,000 immature haploid pollen grains (microspores), each of which has the potential of developing into a haploid plant. This compares favourably to the single egg cell of the ovary. Clapham (1971) was the first to describe the response of barley anthers to *in vitro* culture in which callus was produced. From this initial work complex procedures have now been developed for the production of barley doubled haploids. There are many variations in the protocols of anther culture but the basic method is as follows. Donor plants are normally grown in optimal conditions. Field grown plants are good sources of material, but for all year production of responsive anthers, plants are normally grown at 12°C. Spikes are harvested when the developing pollen grains are at the uninucleate stage. The spikes are then pretreated in Petri dishes at 4°C for 4 weeks (Huang & Sunderland, 1982), the anthers removed and cultured on a nutrient medium containing growth regulators and maltose as a carbohydrate source. The embryos produced are subcultured until green shoots are produced and then cultured on a rooting medium. Rooted plants are subsequently potted (Powell *et al.*, 1991).

The potential of anther culture has not yet been fully realised. One serious problem has been the low numbers of green plants and the high frequency of albino plants produced. Genetic instability, chromosomal abnormalities and somaclonal variation (Larkin & Scowcroft, 1983) have also been found to be problems in plants regenerated from callus (Powell *et al.*, 1984, 1986). These problems have been addressed to a large extent by improvements in culture procedures. For instance, the substitution of sucrose by maltose has significantly increased the number of green plants produced by anther culture (Finnie *et al.*, 1989). Maltose was also found to induce green plant production by promoting embryogenesis, thus bypassing the problematic callus phase, and producing genetically stable plants (Finnie *et al.*, 1991). The production of albino plants is also known to be under genetic control. Using randomly amplified polymorphic DNA (RAPD) markers in conjunction with bulked segregant analysis, Andersen *et al.* (1995) showed that the capacity to produce green plants in barley anther culture could be explained by the action of two genes, one of which is associated with a gene determining spring/winter habit. This is an interesting finding since the most responsive barley cultivars to anther culture are winter types such as Igri (Foroughi-Wehr *et al.*, 1982; Larsen *et al.*, 1991), inferring that genes controlling plant development in winter cultivars may also be active in determining tissue culture response.

### 2.4. *Microspore culture*

Microspore culture is a more recent development than anther culture. Here the pollen grains are separated from their anthers and cultured directly on liquid medium. Depending on the constitution of the culture medium, either callus or embryo development may be induced. These tissues can then be subcultured onto a solid medium for plant regeneration (see Pickering & Devaux, 1992, for more details). Although the technique eliminates any possible negative effects of anther tissues in culture, which have been found to be inhibitory in other species and preclude the development of anther-derived callus, it has not yet proved to be any more efficient than anther culture in the production of barley haploids. The isolation of microspores does, however, provide greater opportunities for cell manipulation such as transformation and *in vitro* selection.

### 2.5. *The* bulbosum *technique*

Interspecific crosses in the Triticeae have frequently resulted in the production of haploids. In most cases, the frequency of haploids is low but in crosses between *H. vulgare* and *H. bulbosum*, high numbers of haploids are produced. Initial work by Kao & Kasha (1969), and confirmed by Kasha & Kao (1970), Subrahmanyam & Kasha (1973), and Bennett *et al.* (1976), showed that haploids in barley were produced after egg fertilisation by progressive elimination of *H. bulbosum* chromosomes during the early stages of embryo development. Haploid plants are obtained by *in vitro* culture of excised embryos. As with anther culture, increased efficiency in haploid production using the *bulbosum* method has involved refinements in parent plant growth conditions, selection of responsive parental types, and *in vitro* embryo culture techniques (reviewed by Pickering & Devaux, 1992). Large numbers of haploids can be produced by this method, but it requires time consuming crossing and embryo culture procedures.

The *bulbosum* technique has in the past been the preferred method of producing doubled haploids in barley, mainly because it can be applied to a wide range of genotypes (Kasha & Reinbergs, 1981) and is not subject to genetic instability found in some anther culture systems (Powell *et al.*, 1984, 1986). There is a genetic component to the success of both methods and each is limited to different degrees by the number of responsive genotypes (Pickering, 1980). In the case of the *bulbosum* technique, the *H. bulbosum* clone used also affects success rates (Simpson *et al.*, 1980). The limitations of anther culture, namely instability, callus formation and albino plant production have been, and continue to be, addressed and it is likely that either this method or microspore culture will become the most efficient haploid production system in barley. With the advent of chromosome painting techniques such as genomic *in situ* hybridisation, it has become apparent that in other crosses where chromosome elimination occurs, e.g., *Solanum tuberosum* × *S. phureja* (Clulow *et al.*, 1991), transfer of chromosome segments can take place. The plants produced therefore may not be *bona fide* haploids but contain some hybridity. Whether this occurs in *H. vulgare* × *H. bulbosum* crosses has not been established, but requires investigation.

## 3. Exploitation of barley haploids: doubled haploids

In barley, as in many other plant species, the main use of haploids is in the production of doubled haploids. Haploidisation followed by chromosome doubling offers the quickest route to homozygosity and, in self-pollinating crops such as barley, this has great potential in plant breeding programmes. Since homozygous lines of barley are true breeding it is possible to maintain a genotype indefinitely. Such "immortal" lines can be repeatedly sampled, a characteristic which can be exploited in a range of genetic studies.

### 3.1. *Doubled haploids in barley breeding*

The genetic improvement of most inbreeding crops such as barley is through the development of true breeding, homozygous lines. These are produced traditionally by the hybridisation of two or more parental lines followed by repeated selfing and selection cycles (Bingham, 1975). Pedigree selection, like other conventional methods in plant breeding, is long term; it can take 10–15 years to produce a barley cultivar by these methods. This time delay is not only inefficient and expensive but makes barley breeding inept in responding quickly to changing demands of agronomy, industry, commerce and society.

A major limitation of pedigree inbreeding is the biasing effect that dominance has in determining the phenotype of early generations (Powell *et al.*, 1986a). In the early stages of pedigree selection virtually all individuals are unique. The unit of selection, the individual, is therefore small, heterozygous and unreplicated, and consequently the data produced are unreliable. In addition, competitive interaction between phenotypes may occur. This of course does not happen in crop conditions, and may influence selection. There is therefore an inability to select desirable genotypes in early generations in pedigree breeding, when progenies are heterozygous and where selection conditions bear little resemblance to field conditions of the crop. Plant breeders are therefore forced to delay intensive selection until sufficient seed is available and progenies approach homozygosity, at which stage several lines will have been eliminated. In order to speed up the development of homozygous lines the technique of single seed descent was developed (Goulden, 1939). Here the same number of generations (usually 5–8) is required to reach a reasonable level of homozygosity but the generation time and the number of lines developed are reduced. This is done by growing single seeds from each plant in restricted growing conditions in a glasshouse (Brimm, 1966). Single seed descent is a fairly straightforward technique, where the number of rounds of recombination is fixed by the number of generations. The products of single seed descent appear to represent a random sample of parental combinations, but as in pedigree breeding there is a risk of selective elimination through competitive interactions. For spring barley, three generations can be produced in a year. However, the advancement of winter barley lines, which require vernalisation, is less rapid. A further disadvantage is the non-random loss of genotypes during the procedure (Riggs & Hayter, 1975).

Doubled haploidy overcomes many of the limitations of conventional plant breeding and single seed descent methods. The main advantages are that homozygosity is achieved in a single generation and the homozygous lines can be quickly bulked allowing early screening for field performance. Doubled haploids can be extracted from any generation of a breeding programme and the breeder can decide on the number of rounds of recombination before homozygosity is attained (Snape & Simpson, 1981; Choo *et al.*, 1985). Several criteria have been identified for the successful and cost-efficient incorporation of doubled haploids into plant breeding. These are: (a) the efficiency of producing haploids and chromosome doubling to produce doubled haploids; (b) the production of large numbers of doubled haploids from a wide range of genotypes; (c) the production of doubled haploids which are genetically stable and of good agronomic performance; and (d) the production of a random sample of parental gametes in the doubled haploid population produced from a cross (Kasha & Reinbergs, 1981; Snape *et al.*, 1986). Given these criteria the *bulbosum* system would appear to be the most suitable method of producing doubled haploids. However, as discussed above, many of the limitations in anther/microspore culture such as genetic stability and range of responsive genotypes have and are being overcome. There remains a problem of distorted segregation of genes in anther/microspore culture-derived doubled haploids (Powell *et al.*, 1986; Foroughi-Wehr & Friedt, 1984; Thompson *et al.*, 1991), but this does not occur to an extent which limits application in plant breeding and anther/microspore culture has now overtaken the *bulbosum* technique as the method of choice in plant breeding. The field performance of barley doubled haploids derived from the *bulbosum* and anther culture techniques has been compared with lines developed from the same crosses using the pedigree, single seed descent and bulk plot methods, and in general there is little difference among the various methods (Reinbergs *et al.*, 1976; Park *et al.*, 1976; Song *et al.*, 1978; Powell *et al.*, 1986b).

#### 3.1.1. *Advantages of doubled haploid in barley breeding*

Doubled haploid production by the *bulbosum* or anther culture method is a time-saving exercise. It is possible to go from crossing parental lines to harvesting replicated field trials in a relatively short time, 24 months. Since doubled haploids are true breeding lines, uniform bulks can be produced in a straightforward manner. Doubled haploid families therefore provide material which can be efficiently screened in replicated field conditions without complications of dominance and competitive interactions. The first barley cultivar to be developed from doubled haploidy was Mingo. This was produced by the *bulbosum* technique in Canada in the late 1970s (Ho & Jones, 1980). The advantages of doubled haploids are now being realised in plant breeding programmes throughout the world and their exploitation in barley breeding is well documented (Kasha & Kao, 1970; Nitzsche & Wenzel, 1977; Kasha & Reinbergs, 1981; Foroughi-Wehr & Friedt, 1984; Wenzel, 1985).

### 3.2. *Doubled haploids in barley genetics*

#### 3.2.1. *Linkage maps*

The production of genetic linkage maps is fundamental to genetic studies in any organism since they allow the phenotype to be related to the genotype. A doubled haploid population produced from an $F_1$ plant is ideal material from which to construct a genetic map: the doubled haploid lines have undergone just one round of recombination and therefore exhibit maximal expression of linkage relationships. Genetic mapping is based on co-segregant analysis of alleles at polymorphic loci (genetic markers). The approach has been greatly facilitated by developments in methods to detect polymorphism, particularly at the molecular level. The genetic markers used in mapping fall into four general classes: (1) morphological; (2) cytological; (3) biochemical; and (4) molecular. Doubled haploids developed from an $F_1$ hybrid will on average segregate in a 1:1 ratio for genetic markers, thus just two genotypic classes are produced. In recent years, several groups have exploited these advantages in producing genetic maps of barley. Principal amongst these is the North American Barley Genome Mapping Project (NABGMP) which incorporates over 60 scientists. The project is based on *bulbosum*-derived doubled haploid populations from three crosses which combine malting quality and high yield traits: Steptoe × Morex; Harrington × TR306; and Harrington × Morex. The Steptoe × Morex cross was the first to be published (Kleinhofs *et al.*, 1993), and is updated regularly in the annual Barley Genetics Newsletter (Eds. M.P. Davis, D.E. Falk & J.D. Franckowiak, North Dakota State University). The map currently contains over 380 loci, these are primarily restriction fragment length polymorphic (RFLPs) loci, but isozyme and morphological markers are also included. RFLP maps have also been produced in Germany using anther culture-derived doubled haploids from interspecific (Vada × *H. spontaneum*) and intraspecific (Igri × Franka) barley crosses (Graner *et al.*, 1991). The German group is currently concentrating on developing the map of winter barley germplasm (Igri × Franka which segregates for various agronomic traits); this map currently contains 273 loci covering a total map distance of 1433 cM (Graner *et al.*, 1993). Maps based on randomly amplified polymorphic DNA (RAPDs) have also been developed (Powell *et al.*, 1992; Giese *et al.*, 1994). The development of linkage maps in barley is now becoming routine and is being accelerated by improvements in the production of doubled haploids and by the development of user friendly and automated DNA analysis technology. It should be stressed that linkage maps created by the *H. bulbosum* and anther culture techniques may not be directly comparable because male and female gametes are being sampled separately. A future objective may be to undertake such a comparison.

### 3.2.2. *QTL analysis*

The development of genetic maps in barley is finding application in the analysis of quantitative trait loci (QTLs). Most characters of interest to plant breeders are quantitative, i.e., controlled by several genes. These traits include yield, days to maturity, plant height, disease resistance, malting quality, etc. Classification of doubled haploid populations into individuals carrying alternative alleles of marker genes allows the effect of that particular marker allele on a given quantitative trait to be assessed. The map distance between a QTL and a marker locus can be determined and the additive effect of the QTL can be determined. The concept of using linked genes to analyse QTLs in barley doubled haploids is not new (Choo *et al.*, 1979; Choo, 1983), but the recent and rapid development of high density linkage maps has opened up the barley genome to more precise genetic analyses. The locations of QTLs with respect to marker genes have been published by the NABGMP, the German group and ourselves (Han & Ullrich 1993; Tinker & Mather, 1993; Backes *et al.*, 1993; Thomas *et al.*, 1993, respectively).

3.2.3. *Bulked segregant analysis*

Recent developments in molecular marker techniques have provided improved methods in the study of both qualitative and quantitative traits in higher plants (Rafalski & Tingey, 1993). One rapid method of detecting markers linked to genes controlling specific traits is to use molecular markers in association with bulked segregant analysis (Michelmore *et al.*, 1991). This approach, based on doubled haploids, has been used successfully in barley to detect markers linked with single genes for disease resistance (Barua *et al.*, 1993) and QTLs controlling milling energy requirement (Chalmers *et al.*, 1993). In dissecting QTLs into their individual genetic components, bulked segregant analysis permits plant material to be chosen with maximal segregation for the quantitative trait. Since quantitative traits are characterised by continuous frequency distributions, made up of a range of individuals segregating for genes controlling the trait, genotypes found in the two extreme tails of the distribution are expected to differ at most of the loci controlling the trait. They are also expected to differ for linked genetic markers. This observation was exploited by Michelmore *et al.* (1991) in bulked segregant analysis. In this technique, DNA from extreme members of the distribution are pooled to form two bulks which are then screened with genetic markers to detect those which show differences between the bulks. Doubled haploid lines developed from $F_1$ hybrids are ideally suited to this type of analysis. An appropriate example here is the development of RAPD markers linked to the genetic control of green/albino plant formation in barley anther culture (Fig. 3). Two loci controlling green/albino plant formation were detected using this technique which account for most of the genetic variation associated with this trait.

3.3. *Transformation*

The latest and perhaps the most exciting application of haploidy in barley is in transformation. The introduction of new traits by gene transfer has become routine for many plant species. Various methods of gene transfer are available but the method used is determined to a large extent by the plant species to be transformed. Methods include the use of *Agrobacterium*, protoplasts, and biolistics (Potrykus, 1991). The most common and efficient transformation system for dicotyledonous plants is *Agrobacterium*-mediated. Monocots

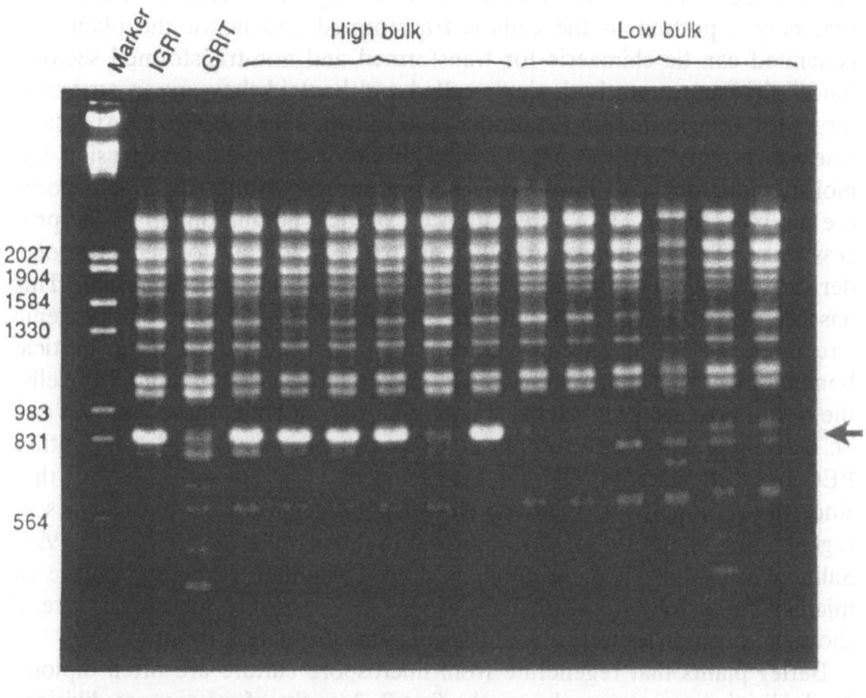


*Figure 3.* **Bulked segregant analysis in a barley doubled haploid population segregating for the capacity to produce high numbers of green regenerants in anther culture. The responsive parent, Igri, is seen to have a RAPD product using primer OPX9–H850 which is not present in the unresponsive parent, Grit. Six of the least responding lines (low bulk) have the genotype of Grit whereas five of the six top responding doubled haploids (high bulk) carry the RAPD marker indicating a strong linkage to a QTL for green plant production.**

such as barley are not natural hosts to the mediating organism, *A. tumefaciens*, and transformation by this route is difficult. However, there has been a report of *Agrobacterium*-mediated transformation in rice (Hiei *et al.*, 1994). Protoplast-mediated transformation of rice and barley has been reported (Datta *et al.*, 1992; Lazzeri *et al.*, 1991, respectively), but regeneration from protoplasts has been troublesome and inefficient. The invention of the particle gun (Klein *et al.*, 1987) has facilitated gene transformation in many of the recalcitrant monocot species and there have been several recent reports of barley transformation using particle gun bombardment of suspensor cells (Mendel *et al.*, 1989), callus tissue (Kartha *et al.*, 1989), endosperm (Knudson & Mueller, 1991) and immature embryos (Ritala *et al.*, 1993, 1994). Immature embryos are currently a popular target in biolistic-mediated transformation, as tissue culture procedures are well-established for the subsequent regeneration into plants. Plants can be regenerated either directly from single cultured embryos or from somatic embryos induced from the scutellar tissues of a cultured embryo. One problem of targeting a multicellular structure is that only a portion of the cells is transformed and hence the plants regenerated can be chimaeric for transformed and non-transformed sectors. Since microspores are both single-celled and haploid they are an attractive target for transformation (Ziauddin *et al.*, 1990; Jähne *et al.*, 1994). Many research groups have attempted transformation of microspores using the biolistic approach but have received little success. Either the microspores are not receptive to DNA integration via particle bombardment or the process causes lethal damage. Another approach is to transform microspore-derived protoplasts but such protoplasts are difficult to regenerate. This has been overcome by first culturing microspores to produce embryogenic structures; 10–30 celled embryos can then be transformed using particle bombardment. Since the embryos contain a relatively small number of cells, the risk of producing chimaeras is reduced. Alternatively, these embryos can be used as a source of protoplasts which can be transformed using either PEG or electroporation DNA delivery systems. The advantage here is that since the protoplasts are derived from embryogenic tissues the chances of regenerating plants are greatly improved (Salmenkallio-Marttila *et al.*, 1994; Salmenkallio-Marttila & Kauppinen, 1995). Another attractive feature of microspore-derived transformation is that it is possible to subject several thousand protoplasts to one simultaneous transformation event.

Barley plants that regenerate from microspore culture are often diploid. Diploidisation may occur during the first 2–3 weeks of microspore division. Spontaneous doubling is a useful mechanism as it precludes the use of chemical doubling agents, such as colchicine which may cause DNA damage. Diploidisation, however, tends to occur before transformation technologies can be applied, e.g., most protoplasts derived from embryonic structures are often already doubled, and this has its drawbacks. It would be more advantageous to transform haploid cells which could then be doubled to produce homozygous, true-breeding transformants. It must be borne in mind that transformation of cereals is in its early days and the efficiency of producing stable transformed plants is a current priority. Consequently much work is focused on successful regeneration systems. Work will however shift in the future to the targeting of haploid cells which have greater potential.

Genes being targeted for barley transformation include genes determining malting quality, e.g., β-glucanase and α-amylase genes (Mannonen *et al.*, 1993), genes for disease and pest resistance, e.g., viral coat protein genes (Wan & Lemaux, 1994), the *Bt* toxin gene (Gatehouse *et al.*, 1991), and genes for abiotic stress tolerance, e.g., constitutively regulated *Sod* and *Per* genes, and herbicide resistance, e.g., the *bar* gene (Thompson *et al.*, 1987).

### 3.4. In vitro *selection*

*In vitro* selection is another area which has particular relevance to haploids. The ability to select mutations is one example. Mutants in barley can be produced in several ways, usually by exposure to ionising radiation treatments, e.g., $\gamma$-rays or chemical mutagens such as sodium azide. The mutants produced are predominantly recessive and therefore difficult to detect in the heterozygous state where their expression is masked by the presence of the dominant allele. There are no such difficulties with haploids such that recessive mutations are more likely to be detected and then fixed in the homozygous state after chromosome doubling. *In vitro* selection of mutants has played an important role in biochemical studies such as amino acid metabolism, carbon and nitrogen metabolism, secondary metabolites, etc. It has also been useful in the selection of novel traits for crop improvement such as herbicide resistance, disease resistance and desired morphological change (Chaleff, 1981; see also Lal & Lal, 1990, for a review of cell selection in cereals). Despite the obvious benefits of haploidy in mutation and *in vitro* selection, there are few examples of their successful application in barley. This is due in part to the fact that researchers of the pure science disciplines have tended to concentrate their efforts on model species such as *Arabidopsis* whereas, in the case of barley breeding, more conventional approaches seem to have sufficed. Mutations produced by treating immature barley haploid embryos with the mutagen N-methyl-N-nitrosourea have been investigated (Pickering & Devaux, 1992), but few useful variants were found. Ye *et al.* (1987) were, however, successful in selecting salt tolerant barley lines by culturing $F_1$ anthers from a cross between a salt tolerant and salt sensitive line on media containing salt ($Na_2SO_4$). Clearly, if mutation and *in vitro* selection are to be pursued in barley, anther/microspore culture has great potential, since it is possible to screen several thousand individuals at the unicellular stage.

## 4. Acknowledgements

We wish to thank our colleague, Dr. Amar Kumar, for his constructive comments in the preparation of the text.

## 5 References

Backes, G., G. Fischbeck, A. Graner & A. Jahoor (1993) Localisation of agronomically important traits in barley. Barley Genet. Newsl. 23: 60–64.

Barua, U.M., K.J. Chalmers, C.A. Hackett, W.T.B. Thomas, W. Powell & R. Waugh (1993) Identification of RAPD markers linked to a *Rhyncosporium secalis* resistance locus in barley using near-isogenic lines and bulked segregant analysis. Heredity 71: 177–184.

Bennett, M.D., R.A. Finch & I.R. Barclay (1976) The time, rate and mechanism of chromosome elimination in *Hordeum* hybrids. Chromosoma 54: 175–200.

Bingham, J. (1975) Winter wheat breeding methods and prospects. J. Agric. Soc. Engl. 136: 65–67.

Brimm, C.A. (1966) A modified pedigree method of selection in soybean. Crop Sci. 6: 220.

Chaleff, R.S. (1981) Variants and mutants. In: D.R. Newth & J.G. Torrey (Eds.), Genetics of Higher Plants: Applications of Cell Culture, pp. 41–95. Cambridge University Press, Cambridge.

Chalmers, K.J., U.M. Barua, C.A. Hackett, W.T.B. Thomas, R. Waugh & W. Powell (1993) Identification of RAPD markers linked to genetic factors controlling the milling energy requirement of barley. Theor. Appl. Genet. 87: 314–320.

Choo, T.M. (1983) Doubled haploids for locating polygenes. Can. J. Genet. Cytol. 25: 425–429.

Choo, T.M., B.R. Christie & E. Reinbergs (1979) Doubled haploids for estimating genetic variances and a scheme for population improvement in self-pollinating crops. Theor. Appl. Genet. 54: 267–271.

Choo, T.M., E. Reinbergs & K.J. Kasha (1985) Use of haploids in breeding barley. Plant Breed. Rev. 3: 219–252.

Clapham, D. (1971) *In vitro* development of callus from the pollen of *Lolium* and *Hordeum*. Z. Pflanzenzücht. 69: 142–155.

Clulow, S.A., M.J. Wilkinson, R. Waugh, E. Baird, M.J. De Maine & W. Powell (1991) Cytological and molecular observations on *Solanum phureja*-induced dihaploid potatoes. Theor. Appl. Genet. 82: 545–551.

Datta, S.K., K. Datta, N. Soltanifar, G. Donn & I. Potrykus (1992) Herbicide-resistant Indica rice plants from IRRI breeding line IR72 after PEG-mediated transformation of protoplasts. Plant Mol. Biol. 20: 619–629.

Fedak, G. (1992) Perspectives on wide crossing in barley. In: L. Munck (Ed.), Barley Genetics VI, pp. 683–699. Munksgaard International, Copenhagen.

Finnie, S.J., A. Dyer, K.J. Chalmers, B.P. Forster & W. Powell (1991) Genetic stability of microspore-derived families of barley: A cytological, biochemical and molecular study. Genome 34: 923–928.

Finnie, S.J., W. Powell & A.F. Dyer (1989) The effects of carbohydrate composition and concentration on anther culture response in barley (*Hordeum vulgare* L.). Plant Breed. 103: 110–118.

Foroughi-Wehr, B., W. Friedt & G. Wenzel (1982) On the genetic improvement of androgenetic haploid formation in *Hordeum vulgare* L. Theor. Appl. Genet. 62: 233–239.

Foroughi-Wehr, B. & W. Friedt (1984) Rapid production of recombinant barley yellow mosaic virus resistant *Hordeum vulgare* lines by anther culture. Theor. Appl. Genet. 67: 377–387.

Gatehouse, A.M.R., V.A. Hilder & J.A. Gatehouse (1991) In: W. Powell, J.R. Hillman & J.A. Raven (Eds), Opportunities and Problems in Plant Biotechnology. Proc. Roy. Soc. Edin. Section B 99: 51–60.

Giese, H., A.G. Holm-Jensen, H. Mathiassen, B. Kjaer, S.K. Rasmussen, H. Bay & J. Jensen (1994) Distribution of RAPD markers on a linkage map of barley. Hereditas 120: 267–273.

Goulden, C.H. (1939) Problems in plant selection. In: R.C. Burnett (Ed.), Proc. 7th Int. Genet. Congress, Edinburgh, pp. 132–133.

Graner, A., E. Bauer, A. Kellermann, S. Kirchner, J.K. Muraya, A. Jahoor & G. Wenzel,, (1993) Progress of RFLP-map construction in winter barley. Barley Genet. Newsl. 23: 53–59.

Graner, A., A. Jahoor, J. Schondelmaier, H. Siedler, K. Pillen, G. Fischbeck, G. Wenzel & R.G. Herrmann (1991) Construction of an RFLP map in barley. Theor. Appl. Genet. 83: 250–256.

Hagberg, A. & G. Hagberg (1980) High frequency of spontaneous haploids in the progeny of an induced mutation in barley. Hereditas 93: 341–343.

Hagberg, A. & G. Hagberg (1987) Production of spontaneously doubled haploids in barley systems with marker genes and the "hap"-gene. Biologisches Zentralblatt 106: 53–58.

Han, F. & S.E. Ullrich (1993) Mapping of quantitative trait loci associated with malting quality in barley. Barley Genet. Newsl. 23: 84–97.

Hansen, A.L., C. Plever, H.C. Pedersen, B. Keimer & S.B. Andersen (1994) Efficient *in vitro* chromosome doubling during *Beta vulgaris* ovule culture. Plant Breed. 112: 89–85.

Hiei, Y., Ohta, S., Komari, T. & T. Kumashiro (1994) Efficient transformation of rice (*Oryza sativa* L.) mediated by *Agrobacterium* and sequence analysis of the boundaries of the T-DNA. Plant J. 6: 271–282.

Ho, K.M. & G.E. Jones (1980) Mingo barley. Can. J. Plant Sci. 60: 279–280.

Huang, B. & N. Sunderland (1982) Temperature-stress pretreatment in barley anther culture. Ann. Bot. 49: 77–88.

Jähne, A., D. Becker, R. Brettschnieder & H. Lörz (1994) Regeneration of transgenic, microspore-derived, fertile barley. Theor. Appl. Genet. 89: 525–533.

Kao, K.N. & K.J. Kasha (1969) Haploidy from interspecific crosses with tetraploid barley. In: R.A. Nilan (Ed.), Barley Genetics II, pp. 82–88. Washington State University Press, Pullman, Wash.

Kartha, K.K., R.N. Chibbar, F. Georges, N. Leung, K. Caswell, E. Kendal & J. Qureshi (1989) Transient expression of chloramphenicol acetyltransferase (CAT) gene in barley cell cultures and immature embryos through microprojectile bombardment. Plant Cell Rep. 8: 429–432.

Kasha, K.J. & K.N. Kao (1970) High frequency haploid production in barley (*Hordeum vulgare* L.). Nature 225: 874–875.

Kasha, K.J. & E. Reinbergs (1981) Recent developments in the production and utilization of haploids in barley. In: M.J.C. Asher, R.P. Ellis, A.M. Hayter & R.N.H. Whitehouse (Eds.), Barley Genetics IV, pp. 655–665. Edinburgh University Press, Edinburgh.

Klein, T.M, E.D. Wolf, R. Wu & J.C. Sanford (1987) High-velocity microprojectiles for delivering nucleic acids into living cells. Nature 327: 70–73.

Kleinhofs, A., A. Kilian, M.A.S. Maroof, R.M. Biyashev, P. Hayes, F.Q. Chen, N. Lapitan, A. Fenwick, T.K. Blake, V. Kanazin, E. Ananiev, L. Dahleen, D. Kudrna, J. Bollinger, S.J. Knapp, B. Liu, M. Sorrells, M. Heun, J.D. Franckowiak, D. Hoffman, R. Skadsen & B.J. Steffenson (1993) A molecular, isozyme and morphological map of the barley (*Hordeum vulgare*) genome. Theor. Appl. Genet. 86: 705–712.

Knudson, S. & M. Mueller (1991) Transformation of the developing barley endosperm by particle bombardment. Planta 185: 330–336.

Lal, R. & S. Lal (1990) Cell selection and long-term high-frequency regeneration of cereals and legumes. In: R. Lal & S. Lal (Eds.), Crop Improvement Utilizing Biotechnology, pp. 195–254. CRC Press Inc., Boca Raton, Florida.

Larkin, P.J. & W.R. Scowcroft (1983) Somaclonal variation and crop improvement. In: T. Kosuge, C.P. Meredith & A. Hollaender (Eds.), Genetic Engineering of Plants: An Agricultural Perspective, pp. 289–314. Plenum Press, New York.

Larsen, E.T., I.K.D. Tuvesson & S.B. Andersen (1991) Nuclear genes affecting percentage of green plants in barley *Hordeum vulgare* L. anther culture. Theor. Appl. Genet. 199: 417–420.

Lazzeri, P.A., R. Brettschneider, R. Luhrs & H. Lorz (1991) Stable transformation of barley via PEG-induced direct DNA uptake into protoplasts. Theor. Appl. Genet. 81: 437–444.

Mannonen, L., U. Kurten, A. Ritala, M. Salmenkallio-Marttila, R. Hannus, K. Aspegren, T. Teeri & V. Kauppinen (1993) Biotechnology for the improvement of malting barley. In: EBC Congress, pp. 85–93.

Mendel, R.R., B. Mueller, J. Schulze, V. Kolesnikov & A. Zelenin (1989) Delivery of foreign genes to intact barley cells by high-velocity microprojectiles. Theor. Appl. Genet. 78: 31–34.

Michelmore, R.W., I. Paran & R.V. Kesseli (1991) Identification of markers linked to disease resistance genes by bulked segregant analysis. A rapid method to detect markers in specific genomic regions by using segregating populations. Proc. Natl. Acad. Sci. U.S.A. 88: 9828–9832.

Mogensen, H.L. (1982) Double fertilisation in barley and the cytological explanation for haploid embryo formation, embryoless caryopses, and ovule abortion. Carlsberg Res. Commun. 47: 313–354.

Nitzsche, W. & G. Wenzel (1977) Haploids in Plant Breeding. Verlag Paul Parey, Berlin.

Park, S.J., E.J. Walsh, E. Reinbergs, L.S.P. Song & K.J. Kasha (1976) Field performance of doubled haploid barley lines in comparison with lines developed by the pedigree and single seed descent methods. Can. J. Plant Sci. 56: 467–474.

Pickering, R.A. (1980) Attempts to overcome partial incompatibility between *Hordeum vulgare* L and *H. bulbosum* L. Euphytica 29: 369–377.

Pickering, R.A. & Devaux, P. (1992) Haploid production: approaches and uses in plant breeding. In: P.R. Shewry (Ed.), Barley: Genetics, Molecular Biology and Biotechnology, pp. 519–547. CAB International, Alden Press Ltd., Oxford.

Potrykus, I. (1991) Gene transfer to plants: Assessment of the published approaches and results. Ann. Rev. Plant Phys. Plant Mol. Biol. 42: 205–225.

Powell, W., U.M. Barua, K.J. Chalmers, W.T.B. Thomas, C.A. Hackett, B.P. Forster & R. Waugh (1992) A foundation linkage map of barley (*Hordeum vulgare* L.) with particular reference to developmentally important genes. Scottish Crop Research Institute Annual Report 1992, pp. 31–33.

Powell, W., E.M. Borrino, M.J. Allison, D.W. Griffiths, M.J.C. Asher & J.M. Dunwell (1986) Genetical analysis of microspore-derived plants of barley (*Hordeum vulgare* L.). Theor. Appl. Genet. 72: 619–626.

Powell, W., P.D.S. Caligari & J.M. Dunwell (1986b) Field performance of lines derived from haploid and diploid tissues of *Hordeum vulgare*. Theor. Appl. Genet. 72: 458–465.

Powell, W., P.D.S. Caligari & W.T.B. Thomas (1986a) Comparison of spring barley lines produced by single seed descent, pedigree inbreeding and doubled haploidy. Plant Breed. 97: 138–146.

Powell, W., B.P. Forster, W.T.B. Thomas, R. Waugh, K.J. Chalmers, U. Barua, L. Hakim, G.R. Young, M. Macaulay, C.A. Hackett & J. McNicol (1991) Doubled haploids: their role in the location and analysis of polygenically controlled traits in barley. Scottish Crop Research Institute Annual Report 1991, pp. 36–40.

Powell, W., A.M. Hayter, W. Wood, J.M. Dunwell & B. Huang (1984) Variation in the agronomic characters of microspore-derived plants of *Hordeum vulgare* cv. Sabarlis. Heredity 52: 13–23.

Rafalski, J.A. & S.V. Tingey (1993) Genetic diagnostics in plant breeding, RAPDs, microsatellites and machines. Trends Genet. 9(8): 275–280.

Rasmusson, D.C. (1992) Barley breeding at present and in the future. In: L. Munck (Ed.), Barley Genetics VI, Vol. II, pp. 865–878. Munksgaard International, Copenhagen.

Reinbergs, E., S.J. Park & L.S.P. Song (1976) Early identification of superior barley crosses by doubled haploid technique. Z. Pflanzenzüchtg. 76: 215–224.

Riggs, T.J. & A.M. Hayter (1975) Practical aspects of the single seed descent method in barley breeding. In: H. Gaul (Ed.), Barley Genetics III. Proc. 3rd Int. Barley Genet. Symp., Garching, 1975, pp. 708–717. Verlag Karl Thiemig, Munich.

Ritala, A., K. Aspegren, U. Kurten, M. Salmenkallio-Marttila, L. Mannonen, R. Hannus, V. Kauppinen, T.H. Teeri & T. Enari (1994) Fertile transgenic barley by particle bombardment of immature embryos. Plant Mol. Biol. 24: 317–325.

Ritala, A., L. Mannonen, K. Aspegren, M. Salmenkallio-Marttila, U. Kurten, R. Hannus, J. Mendez Lozano, T.H. Teeri & V. Kauppinen (1993) Stable transformation of barley tissue culture by particle bombardment. Plant Cell Rep. 12: 435–440.

Salmenkallio-Marttila, M.U. Kurten & V. Kauppinen (1994) Culture conditions for efficient induction of green plants from isolated microspores of barley (*Hordeum vulgare* L.). Plant Cell Tiss. Organ Cult. (in press).

Salmenkallio-Marttila, M. & V. Kauppinen (1995) Efficient regeneration of fertile plants from protoplasts isolated from microspore culture of barley (*Hordeum vulgare* L.). Plant Cell Rep. 14: 253–256.

San Noeum, L.H. (1976) Haploïdes d'*Hordeum vulgare* L. par culture *in vitro* d'ovaires non fecondés. Ann. Amélior. Plant. 26: 751–783.

Simpson, E., J.W. Snape & R.A. Finch (1980) Variation between *Hordeum bulbosum* genotypes in their ability to produce haploids in barley, *Hordeum vulgare*. Z. Pflanzenzücht. 85: 205–211.

Snape, J.W. & E. Simpson (1981) The genetic expectations of doubled haploid lines derived from different filial generations. Theor. Appl. Genet. 60: 123–128.

Snape, J.W., E. Simpson, B.B. Parker, W. Friedt & B. Foroughi-Wehr (1986) Criteria for the selection and use of doubled haploid systems in cereal breeding programmes. In: W. Horn, C.J. Jensen, W. Odenbach & O. Schneider (Eds.), Genetic Manipulation in Plant Breeding, pp. 217–229. De Gruyter, Berlin.

Song, L.S.P., S.J. Park, E. Reinbergs, T.M. Choo & K.J. Kasha (1978) Doubled haploid vs. the bulk plot method for production of homozygous lines in barley. Z. Pflanzenzüchtg. 81: 271–280.

Subrahmanyam, N.C. & K.J. Kasha (1973) Selective chromosomal elimination during haploid formation in barley following interspecific hybridisation. Chromosoma 42: 111–125.

Thomas, W.T.B., M.J.C. Asher, J.S. Swanston & C.E. Thomas (1987) The transfer of resistance to powdery mildew from *Hordeum spontaneum* to the spring barley cv. Golden Promise. In: M.L. Jorna & L.A.J. Slootmaker (Eds.), Proceedings of the Conference of Cereal Section of EUCARPIA, pp. 200–204. Pudoc, Wageningen.

Thomas, W.T.B., W. Powell, K.J. Chalmers, U.M. Barua, J.S. Swanston, R. Waugh, B.P. Forster, R.P. Ellis, P. Jack & V. Lea (1993) Quantitative trait loci for yield and quality components in spring barley. Aspects Appl. Biol. 36: 125–134.

Thompson, C.J., N.R. Novva, R. Tizard, R. Crameri, J.E. Davies, M. Lauwereys & J. Botterman (1987) Characterization of the herbicide-resistance gene *bar* from *Streptomyces hygroscopicus*. Bio/Technology 6: 2519–2523.

Thompson, D.M., K. Chalmers, R. Waugh, B.P. Forster, W.T.B. Thomas, P.D.S. Caligari & W. Powell (1991) The inheritance of genetic markers in microspore-derived plants of barley *Hordeum vulgare* L. Theor. Appl. Genet. 81: 487–492.

Tinker, N.A. & D.E. Mather (1993) Main effects of quantitative trait loci in Harrington/TR306 two-row barley. Barley Genet. Newsl. 23: 72–78.

Von Bothmer, R. (1992) The wild species of *Hordeum*: Relationships and potential use for improvement of cultivated barley. In: P.R. Shewry (Ed.), Barley: Genetics, Molecular Biology and Biotechnology, pp. 3–18. CAB International, Alden Press Ltd., Oxford.

Wan, Y. & P.G. Lemaux (1994) Generation of large numbers of independently transformed fertile barley plants. Plant Physiol. 104: 37–48.

Wenzel, G. (1985) Strategies in unconventional breeding for disease resistance. Ann. Rev. Phytopath. 23: 149–172.

Ye, J.M, Kao, K.N., Harvey, B.L. & B.G. Rossnagel (1987) Screening salt-tolerant barley genotypes via $F_1$ anther culture in salt stress media. Theor. Appl. Genet. 74: 426–429.

Yean-San, L.H. (1987) "Gynogenèse *in vitro*". Variabilité des haploides doubles issus d'androgenèse, de gynogenèse *in vitro* et de croisement interspecifique chez *Hordeum vulgare* L. Ph.D. Thesis, University of Paris, Paris.

Ziauddin, A., E. Simion & K.J. Kasha (1990) Improved plant regeneration from shed microspore culture in barley (*Hordeum vulgare* L.) cv. Igri. Plant Cell Rep. 9: 69–72.

Zohary, D. & M. Hopf (1988) Domestication of Plants in the Old World. Clarendon, Oxford.

# 5. Haploidy in triticale

PÄIVI H. RYÖPPY

**Contents**



## 1. Introduction

Triticale *Triticosecale* (Wittmack), often called the first man-made cereal, is a cross between wheat (*Triticum*) and rye (*Secale cereale*). Depending on the ploidy level triticales can be divided into three main groups:

1. Tetraploids: crosses between diploid wheat and rye;
2. Hexaploids: crosses between tetraploid wheat – usually macaroni wheat *T. durum* – and rye;
3. Octoploids: crosses between hexaploid wheat – usually bread wheat *T. aestivum* – and rye.

Triticales of different groups can be crossed with each other or backcrossed to the parents so that secondary types are formed. In addition to this, wheat has all kinds of substitution, translocation and addition lines, in which one or more rye chromosomes are present. This makes it rather difficult to draw the line between wheat and triticale and thus the Second International Triticale Symposium has suggested that triticale should be included in the genus *Triticum* and the different groups be named as species (Fig. 1) (CIMMYT, 1991).

Triticale is rather young as a species. It has been known for about 120 years (Wilson, 1876) and due to its hybrid nature, there have been many problems in developing it into a cultivated species. One problem has been the instability of the plant, especially at the octoploid level. However, the octoploids as well as the tetraploids can be used as bridges, when transferring genetic material to the hexaploids. Today octoploid varieties also exist. In other "normal" self-breeding species, obtaining homozygous lines after every cross has been expensive and time consuming, whereas in triticale it has been

*S.M. Jain, S.K. Sopory & R.E. Veilleux (eds.)*, In Vitro *Haploid Production in Higher Plants, Vol. 4*, 117–131.

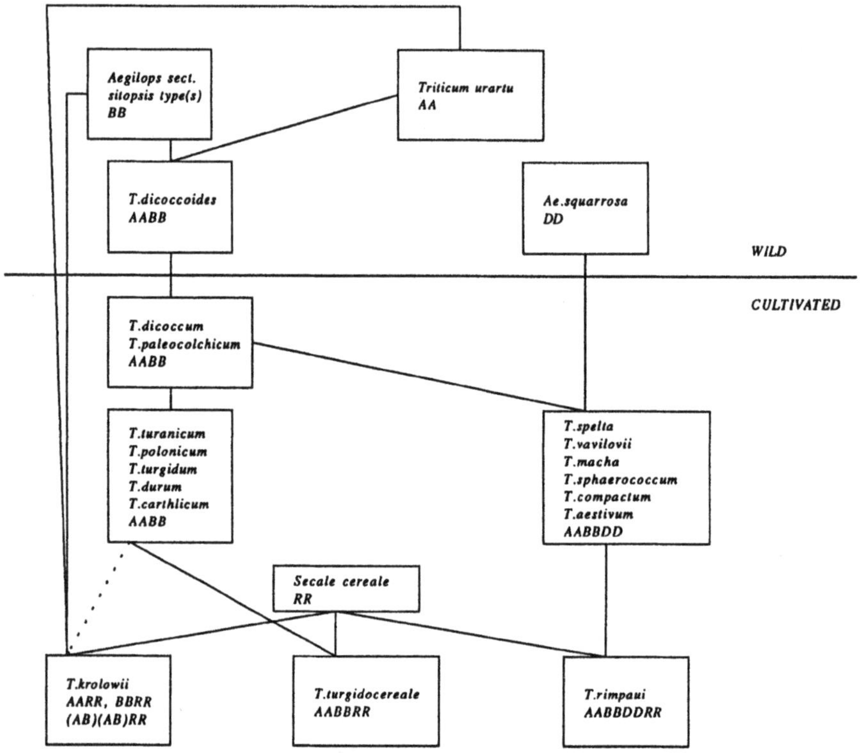


*Figure 1.* The evolution of triticale. Modified from Miller (1987) according the suggestions of the 2nd International Triticale Symposium (CIMMYT, 1991).

twice as difficult. Not only the time needed to make a pure line from a $F_1$-hybrid but also the possibility of genetic instability has to be taken into consideration. An effective method to produce homozygous lines through haploid plants is desirable.

## 2. Production of haploids

The two main techniques of producing haploids are through chromosome elimination in the zygote and through anther or microspore culture. In barley and potato chromosome elimination has been effective, but attempts to extend the method to wheat and triticale have usually failed due to lack of a suitable pollinator. Some success has been achieved in wheat, when pollinated by *Hordeum bulbosum*, but only with some genotypes (Snape *et al.*, 1980). Attempts to pollinate triticale with *H. bulbosum* failed (Kasha, 1986). A better pollinator could be maize (*Zea mays*), which can be used successfully in extracting haploids of hexaploid wheat (Laurie & Bennett, 1988) and, recently, tetraploid wheat (O'Donoughue & Bennett, 1994) or sorghum (Ohkawa *et al.*, 1992).

Wheat haploids can be produced by culturing isolated microspores (Due Tuvesson & Öhlund, 1993), but attempts to culture isolated microspores of triticale have mostly failed and it has been suggested that some factors released from the anther affect pollen development *in vitro* (Keller, 1991). The only technique by which triticale haploids have been produced so far is anther culture. Anther culture of hexaploid and octoploid triticale has been used successfully for more than twenty years in several countries including China (Wang *et al.*, 1973), France (Bernard, 1980), Canada (Chien & Kao, 1983) and Finland (Ryöppy-Ekbom & Tigerstedt, 1984). For tetraploid triticale the first haploids were produced much more recently (Lehman, 1989).

## 3. Factors affecting anther culture response

Several factors are known to affect the anther culture response such as the donor plant, pretreatments, the developmental stage of the pollen, culture media and cultural conditions.

### 3.1. *The donor plant*

The genetic composition of the donor plant is a major factor affecting anther culture response. In fact variation among genotypes (even among spikes within the same variety) can be much greater than variation among treatments and this makes comparisons among treatments difficult (Chien & Kao, 1983; Ryöppy *et al.*, 1986b). The winter types often respond better than spring types although large variation occurs within the groups (Ryöppy-Ekbom & Tigerstedt, 1984). Both the ability to form embryos from microspores and the regeneration of these to plantlets are genetically controlled. The control of embryo formation shows mainly nuclear inheritance, with predominantly additive gene action, while the control of the regeneration process is more complex. There is also a strong environmental influence on the regeneration capacity and the formation of green shoots instead of albinos (Charmet & Bernard, 1984). This leads to the hypothesis that doubled haploids could be better donors in anther culture than others. This is not the case. It has been shown in both triticale (Charmet & Bernard, 1984) and in wheat (Lazar *et al.*, 1984) that the parental lines often outperform their doubled haploid progeny.

Field-grown material is often considered superior over greenhouse-grown material, but if field grown material is used, there are always unpredictable factors that have to be taken in consideration. The weather is one. In Finland the days are long, 16–19 h, during the vegetative growth period in May and early June, but the temperature, especially at nights, can be low. If a cool period follows a warmer one, when the pollen is at the uninucleate stage, the results can be good; 10–15% of anthers may produce embryos in several genotypes under these conditions (Ryöppy, unpublished data). However, under field conditions many factors interact and the results may be unpredictable. This has been shown especially in wheat (Ouyang *et al.*, 1983; Lazar *et al.*, 1984; Bjørnstad *et al.*, 1989).

Another little discussed problem with field-grown material is the use of pesticides. If the vegetation has been treated with phenoxy herbicides prior to anthesis, it may affect the anther culture response. In an experiment in the greenhouse some mother plants were treated with 2,4–D 3–15 days before the spikes were collected for anther culture. The results showed that the treated anthers gave fewer embryos than the control and embryos from treated anthers regenerated only albinos, while the embryos from the control anthers gave both green and albino plants (Ryöppy *et al.*, 1986b; Ryöppy, unpublished data).

### 3.2. *Pretreatments*

Different pretreatments have a great effect on anther culture. A temperature shock at +5°C for various periods of time has proven effective. Lukjanjuk & Ignatova (1986) claimed that a period of 7–15 days was optimal. However, the duration of the temperature shock seems to interact with the genotype. For some genotypes a period as long as 21 days has been optimal (Ryöppy *et al.*, 1986b). When the spikes are kept in nutrient media instead of water prior to anther culture, the cold shock can be more effective. For example three days in $N_6$ media gave 8.7% response while in water the percentage was only 1.8% (Sun *et al.*, 1980). In this respect triticale is different from wheat, in which it has been shown that pretreatment in nutrient medium can lower the response (Marsolais *et al.*, 1984).

### 3.3. *Pollen developmental stage*

Anthers with pollen at the middle to late uninucleate stage can be used to produce haploid plants (Fig. 2). The optimal developmental stage to start the anther culture is the late uninucleate stage and the development then follows the B-pathway according to Sunderland's (1974) classification (Lukjanjuk & Ignatova, 1986). In this type of development two equal, diffuse vegetative-type nuclei are formed. One of the two nuclei undergoes limited divisions only after a short interphase, and a number of free nuclei are formed. These fail to participate in the final product. Division of the second nucleus is delayed for a relatively long period, and it is from this nucleus that the pollen callus ultimately arises. Its first division is accompanied by wall formation and thereafter all divisions are normal (Sunderland, 1974).

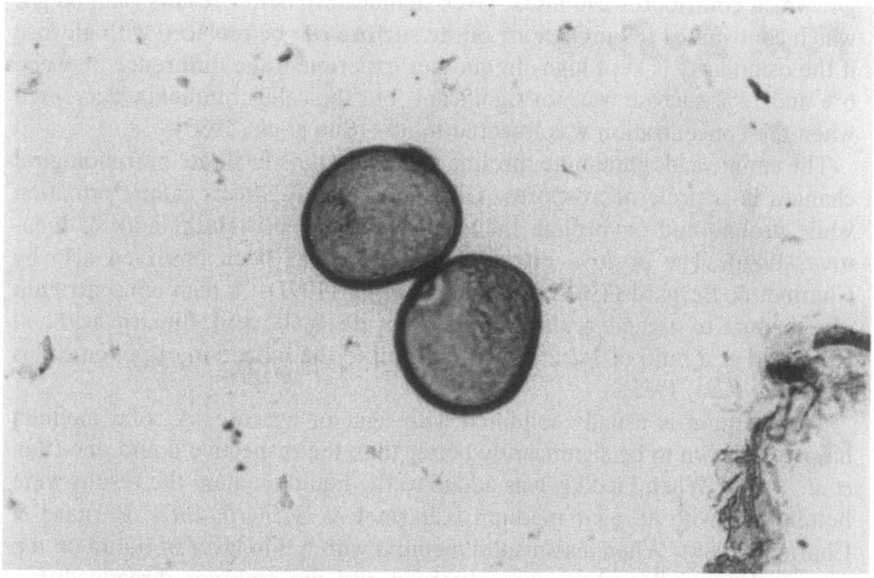

*Figure 2.* Triticale pollen at the uninucleate stage. Aceto-orcein staining.

### 3.4. *Culture medium*

The culture medium has a significant role in callus and/or embryo formation. The most often used basic medium has probably been $N_6$ (Chu *et al.*, 1975), but several others like B5 (Gamborg *et al.*, 1968), K1 (Kao, 1981) and KM (Kao & Michayluk, 1975) have been used as well (Sun *et al.*, 1980; Chien & Kao, 1983; Lukjanjuk & Ignatova, 1986). MS medium (Murashige & Skoog, 1962) has been shown to be significantly less effective than others (Sun *et al.*, 1980).

An auxin such as 2,4–D is essential for callus proliferation, but its concentration can vary from 0.5 to 10 mg/l (2–40 μM) without any significant difference (Sun *et al.*, 1980). The optimal concentration is probably dependent on the genotype of the donor plant. Sometimes sodium and potassium salts of 2,4–D can be used (Lukjanjuk & Ignatova, 1986). In one experiment 2,4–D has been replaced with 0.75–1 mg/l (4–5.5 μM) NAA with good results (Darvey *et al.*, 1991).

The role of cytokinins is more complicated. It is possible they have no effect on embryo formation from microspores (Lukjanjuk & Ignatova, 1986) or they promote pollen callus formation (Chien & Kao, 1983). At least they seem to inhibit callus proliferation from the filament (Sun *et al.*, 1980; Chien & Kao, 1983).

Sucrose concentration is usually high, 6–12%, during the initial culture phase. Chien & Kao (1983) have shown that a sucrose concentration of 100 g/l, which contributes 290 mOs, gives significantly better results than 60 g/l, which contributes 180 mOs. Part of the sucrose can be replaced with glucose if the osmolarity is kept high. In another experiment the difference between 6% and 12% sucrose was not significant, but the callus frequency decreased when the concentration was lowered to 3% (Sun *et al.*, 1980).

The amino acids glutamine, proline and oxyproline facilitate morphological changes in triticale microspores. Glutamine mainly affects callus formation while proline and oxyproline influence embryogenesis (Lukjanjuk & Ignatova, 1986). The positive effect of glutamine has been observed also by Charmet & Bernard (1984) and Darvey *et al.* (1991). A high concentration of a mixture of organic acids (sodium pyruvate, malic acid, fumaric acid and citric acid at a ratio of 1:2:2:2) was harmful to the induction of pollen callus (Chien & Kao, 1983).

The medium is usually solidified with agar or agarose. A solid medium has been shown to be significantly better than the respective liquid one (Sun *et al.*, 1980). When Ficoll® was added to the liquid medium the results were better than with an agar medium (Charmet & Bernard, 1984; Bernard & Charmet, 1985). When a semisolid medium with a thin layer of liquid on top was used, no callus phase was observed and the embryos directly differentiated shoots and roots (Darvey *et al.*, 1991).

### 3.5. *Physical factors*

In Bernard's experiments (Bernard, 1980; Bernard & Charmet, 1985) light had a marked impact on *in vitro* androgenesis; anther culture in the dark allowed a 2.25-fold increase in embryo production compared to light conditions, while Sun *et al.* (1980) observed no difference between light and dark. In my own experiments, weak light – less than 50 $\mu E/m^2/s$ – seemed to be better than complete darkness, although the difference was not significant (Ryöppy, unpublished data).

A high temperature, 27 ± 2°C, in the initiation phase of the culture has promoted callus and/or embryo formation, but for the regeneration phase the temperature must be lowered below 25°C. High temperatures during regeneration increased the frequency of albinos (Bernard, 1980; Ryöppy *et al.*, 1986a).

## 4. Regeneration

The first embryos and/or calli can be formed after about three weeks in culture (Darvey *et al.*, 1991), but normally it takes 5 to 7 weeks (Charmet & Bernard, 1984; Ryöppy *et al.*, 1986b). The "right kind" of callus, which originates from the pollen grains is dense, compact and white or cream

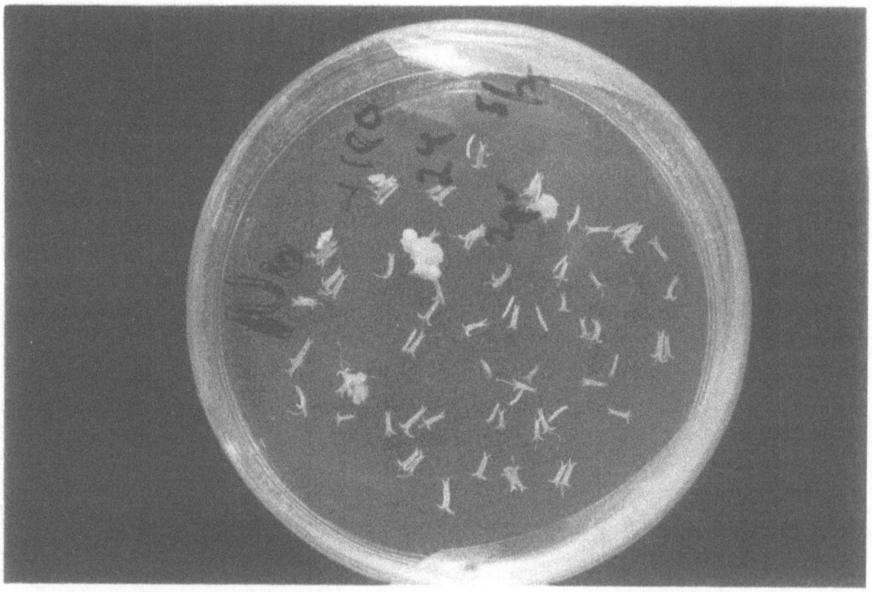

*Figure 3.* Callus and embryos growing from anthers (Petri dish Ø 9 cm).

colored (Fig. 3). Callus formed in less than three weeks usually initiates from the filament and is friable, watery, almost colorless and dies on further cultivation (Chien & Kao, 1983; Lukjanjuk & Ignatova, 1986). Embryos and/or calli formed later than eight weeks in culture usually give rise to albino plants (Ryöppy, unpublished data).The transferred calli should be about 1–2 mm in diameter (Chien & Kao, 1983).

In the regeneration medium 2,4–D is often replaced with low concentrations of other auxins like NAA, 0.5–2.0 mg/l (2.5–10 μM) (Sun *et al.*, 1980; Ryöppy *et al.*, 1986b) or at least the concentration is significantly lowered (Chien & Kao, 1983). Cytokinins like kinetin or zeatin riboside, usually 1 mg/l (5 μM and 3 μM, respectively), are used to promote shoot formation (Sun *et al.*, 1980; Chien & Kao, 1983; Ryöppy *et al.*, 1986a). Sometimes medium without growth regulators can be used (Chien & Kao, 1983). The sucrose concentration should be lowered to a "more normal" level, i.e., approximately 2%.

Regeneration of the callus can be observed within a few weeks after transfer (Charmet & Bernard, 1984; Ryöppy *et al.*, 1986b) (Fig. 4), whereas regeneration of embryos is considerably faster, about one week (Lukjanjuk & Ignatova, 1986).

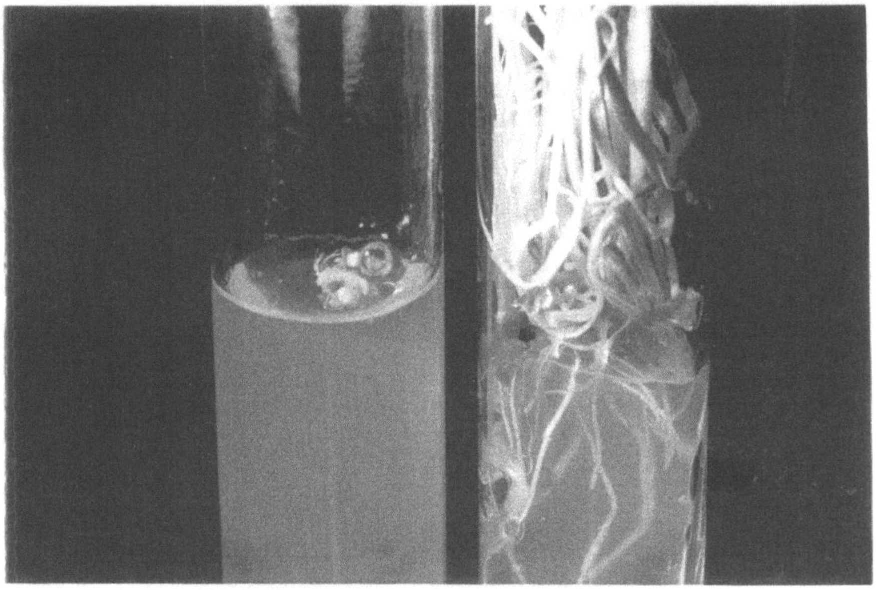

*Figure 4.* Regeneration of triticale callus. On the left shoots proliferating from callus, on the right both green and albino plantlets differentiating from the same callus.

## 5. Chromosome doubling

Colchicine is the most commonly used chromosome doubling agent. In comparison with nitrous oxide, colchicine + DMSO was twice as effective for doubling haploid barley (Subrahmanyam & Kasha, 1975).

Because triticale is an amphiploid between wheat and rye and the original crosses are usually sterile, several methods to double the chromosomes of these hybrids have been developed. Colchicine treatment can be done to the tillers or to the roots at various developmental stages. Beatty *et al.* (1976) cut the main tiller 2 cm from the crown and placed a vial containing 0.1% colchicine in aqueous 2.0% DMSO over the cut stem for several days. Thomas & Kaltsikes (1977) injected 0.03% colchicine solution into the boot of triticale × wheat hybrids 120 h prior to first metaphase of meiosis causing a 40% reduction in chromosome pairing. Linde-Laursen (1973) compared the "tiller method" with "via-the-roots" method and found the latter more effective. In "via-the-root" method the rinsed roots were cut under water to a length of about 3 cm and the plant was allowed to absorb 0.25% aqueous colchicine for 4 h, whereafter the roots were rinsed and the plant repotted. This resulted in 69% fertile, 25% sterile and 6% dead compared with 41% fertile, 49% sterile and 10% dead plants for the tiller method.

Methods developed specially to double haploid triticale from anther culture

*Figure 5.* Colchicine treatment of haploid triticale plantlets.

have been few. Sun *et al.* (1980) let the haploids pass the winter in the field and early in the next year the roots were immersed in 0.025% colchicine for 4 days for chromosome doubling.

A simple colchicine treatment can be done prior to transferring plants to soil. When the plants are removed from the culture vial, the roots can be washed free of agar and cut to a length of 1–2 cm. The plant can then be placed in 0.1% colchicine (aqueous solution) for 4 hours at room temperature such that the shoot tip is also in the liquid (Fig. 5). After the treatment colchicine can be washed away with water and the plant potted. The survival rate has been over 90% and about two thirds of the haploids had doubled. The root tips that were cut off can be used for counting chromosomes (Ryöppy, unpublished data).

## 6. Variation among the regenerated plants

### 6.1. *Chromosomes*

Not all plants regenerated from anther culture are haploid. Sometimes other tissues than pollen start to grow, although callus arising from the filament is usually unable to develop further (Lukjanjuk & Ignatova, 1986). Spontan-

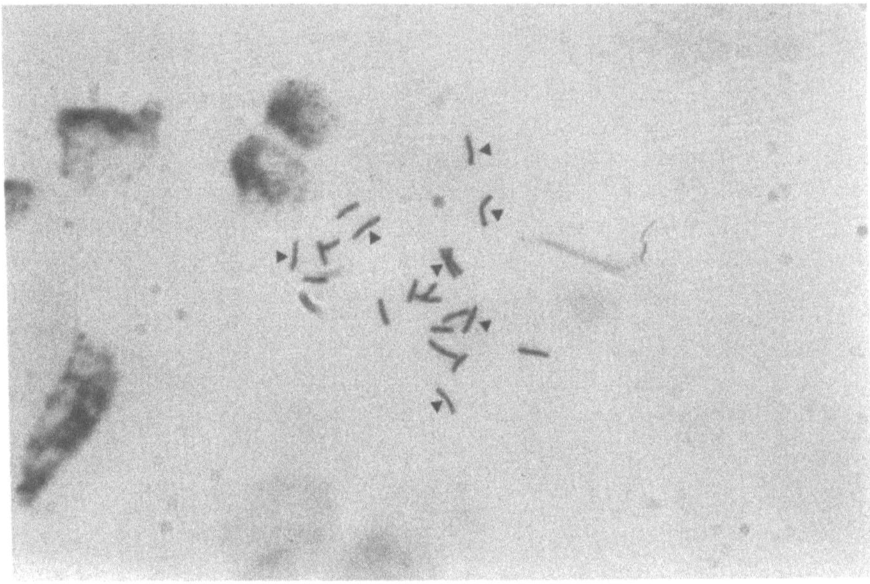

*Figure 6.* Hexaploid triticale ($2n = 42$): haploid chromosomes $n = 21$. The seven rye chromosomes (marked with arrows) can be recognized by their telomeric heterochromatin.

eous diploidization may occur especially if a callus phase is involved, such tissue can be cytologically unstable and give rise to aneuploids.

Charmet (1985) found that, of 408 pollen-derived plants of hexaploid triticale ($2n = 42$), 297 (72.8%) were "haploids", 157 of which were aneuploids with $n = 18$–24; 110 plants (27%) were "diploids", of which 70 were aneuploids with $2n = 34$–45 and one plant was tetraploid (0.2%). Lukjanjuk & Ignatova (1986) found most of their regenerated plants were haploid, but diploids, aneuploids and plants with a higher ploidy level were also found. My own experiences have been similar: about 70% haploids (Fig. 6), 15% diploids and 15% aneuploids (Ryöppy, unpublished data).

The situation is similar in hexaploid triticale × wheat hybrids. About two thirds of microspore-derived plants have been haploid with chromosome number ranging from 17 to 27 and one third were diploid ($2n = 38$–52) (Wang & Hu, 1985). These lines have been shown to be relatively stable cytologically over several generations of selfing. However the rye chromosomes appear to have lost some of their heterochromatin, which made it difficult to establish their continued presence (Wang *et al.*, 1993).

There are two main origins of *in vitro* androgenesis through diploid plants. The first is spontaneous doubling of haploid microspores and the second is androgenic development of unreduced gametes. Aneuploids can arise either from aneuploid microspores or from cytological instability during culture

ture (Charmet, 1985). For triticale it seems that the major cause of aneuploidy preexists in the donor plant (Charmet *et al.*, 1986).

### 6.2. *Albinos*

In cereal tissue culture a serious problem has been the occurrence of albinos. It is typical for half of the regenerated plants to be albino. It is possible to reduce the number of albinos to some extent by pretreatments, culture media and cultural conditions as discussed earlier in this chapter, but the tendency to produce albinos seems to be dependent on the genotype of the mother plant (Ryöppy *et al.*, 1986a,b). The same callus can produce both green and albino shoots (Fig. 4). If chromosome numbers are compared, it seems that there are more haploids than diploids among the albinos whereas the ratio of diploids/haploids is higher among the green plants (Ryöppy, unpublished data).

The reason for albinism is not clear. In wheat and barley Day & Ellis (1984, 1985) reported that albinos have deletions in their chloroplast DNA. In rice it has been shown that the albinos had only about 1.6% chlorophyll a and 2.3% chlorophyll b compared to green plants, and the 23S and 16S ribosomal RNAs were missing (Wang *et al.*, 1978).

### 6.3. *Somaclonal variation*

Among the regenerated plants considerable variation can be observed in the height and habit of the plants, tillering ability, size of the awn, straw thickness, etc. (Lukjanjuk & Ignatova, 1986). Even plants regenerated from the same anther can exhibit wide variation (Fig. 7) (Ryöppy, unpublished data). This is understandable, if we remember that each anther contains hundreds or even thousands of microspores and thus the plants can have different genetic backgrounds.

But even plants regenerated from the same callus can show different properties. When high molecular weight glutenins (HMW glutenins) of doubled haploid lines were studied, such variation was observed. The lines were stable from one year to another, but between lines the differences in the HMW glutenin composition could not be explained by preexisting variation. However, the donor plant may be responsible for this kind of variation: some mother plants produced more stable progeny than others. Even differences from year to year occur (Ryöppy, 1991). A similar situation has been reported in wheat, when different properties were studied (Ryan *et al.*, 1987).

## 7. Field performance of doubled haploids

In the Finnish experiments most doubled haploid lines have been too late for the Finnish climatic conditions. When fifteen hexaploid spring types were

*Figure 7.* Variation among a doubled haploid progeny produced from the same anther. Three from the left have been regenerated from the same callus while the fourth represents another callus = microspore.

studied in field experiments, only two were early enough to be studied further. Surprisingly these two were earlier than the standard wheat variety "Ruso" and much earlier than the parental lines (Ryöppy, unpublished data).

When androgenic doubled haploids were compared to single seed descent (SSD) lines, SSD did not produce a greater range of recombinants although the SSD method theoretically provides more opportunity for recombination. Thus anther culture provides an efficient method for producing high yielding homozygous lines from $F_1$ hybrids of triticale in a relatively short time (Charmet & Branlard, 1985).

## 8. Conclusions

Triticale is an amphiploid derived from a self-pollinator, wheat, and a wind-pollinated, cross-fertilizer, rye. Because it is a combination of divergent species, today's triticales are prone to genetic instability. Anther culture is a valuable tool for creating stability through completely homozygous doubled haploid plants. On the other hand, approximately 10% of anther-derived plants of wheat and wheat hybrids show aneuploidy, including nullisomy, monosomy, trisomy, etc. (Hu, 1989). The frequency of these aneuploids is higher than that obtained from conventional methods. However, anther culture may still be an easier and more efficient method in chromosome engineering than conventional methods.

## 9. References

Beatty, K.D., E.A. Rupert & N. Dehgan (1976) Doubling chromosome numbers of wheat-rye amphiploids with colchicine, DMSO, and cold treatment. Wheat Information Service 43: 10–12.

Bernard, S. (1980) *In vitro* androgenesis in hexaploid triticale: determination of physical conditions increasing embryoid and green plant production. Z. Pflanzenzüchtg. 85: 308–321.

Bernard, S. & G. Charmet (1985) Improvement of haploid production through *in vitro* anther culture in hexaploid triticale. In: Genetics and Breeding of Triticale, EUCARPIA Meeting, Clermont-Ferrand (France), 2–5 July 1984, pp. 306–316. INRA, Paris.

Bjørnstad, Å., H.-G. Opsahl-Ferstad & M. Aasmo (1989) Effects of donor plant environment and light during incubation on anther cultures of some spring wheat (*Triticum aestivum*) cultivars. Plant Cell Tiss. Organ Cult. 17: 27–37.

Charmet, G. (1985) Chromosome structure of androgenetic plants in hexaploid triticale. In: Genetics and Breeding of Triticale, EUCARPIA Meeting, Clermont-Ferrand (France), 2–5 July 1984, pp. 317–328. INRA, Paris.

Charmet, G. & S. Bernard (1984) Diallel analysis of androgenetic plant production in hexaploid triticale (*X. triticosecale*, Wittmack). Theor. Appl. Genet. 69: 55–61.

Charmet, G., S. Bernard & M. Bernard (1986) Origin of aneuploid plants obtained by anther culture in triticale. Can. J. Genet. Cytol. 28: 444–452.

Charmet, G. & G. Branlard (1985) A comparison of androgenetic doubled-haploid, and single seed descent lines in triticale. Theor. Appl. Genet. 71: 193–200.

Chien, Y.C. & K.N. Kao (1983) Effects of osmolality, cytokinin and organic acids on pollen callus formation in triticale anthers. Can. J. Bot. 61: 639–641.

Chu Chih-ching, Wang Ching-chu, Sun Ching-san, Hsu Chen, Yin Kwang-chu, Chu Chih-yin & Bi Feng-yun (1975) Establishment of an effective medium for anther culture of rice through comparative experiments on the nitrogen sources. Scientia Sinica 18: 659–668.

CIMMYT (1991) Business session. In: Proceedings of the Second International Triticale Symposium, Passo Fundo, Rio Grande do Sul, Brazil 1–5 Oct. 1990, pp. 722–723. CIMMYT, Mexico, D.F.

Darvey, N.L., J.A. Fazal, F. Barbera, G. Ancora & A. Rimes (1991) A modified anther culture methodology for increasing embryoid production in wheat and triticale. In: Proceedings of the Second International Triticale Symposium, Passo Fundo, Rio Grande do Sul, Brazil 1–5 Oct. 1990, pp. 314–319. CIMMYT, Mexico, D.F.

Day, A. & T.H.N. Ellis (1984) Chloroplast DNA deletions associated with wheat plants regenerated from pollen: possible basis for maternal inheritance of chloroplasts. Cell 39: 359–368.

Day, A. & T.H.N. Ellis (1985) Deleted forms of plastid DNA in albino plants from cereal anther culture. Curr. Genet. 9: 671–678.

Due Tuvesson, I.K. & R.C.V. Öhlund (1993) Plant regeneration through culture of isolated microspores of *Triticum aestivum* L. Plant Cell Tiss. Organ Cult. 34: 163–167.

Gamborg, O.L., R.A. Miller & K. Ojima (1968) Nutrient requirements of suspension cultures of soybean root cells. Exp. Cell Res. 50: 151–158.

Hu, H. (1989) Chromosome engineering by anther culture. In: A. Mujeeb-Kazi & L.A. Sitch (Eds.), Review of Advances in Plant Biotechnology, 1985–88: 2nd International Symposium on Genetic Manipulation in Crops, Mexico, D.F. CIMMYT, Mexico D.F./IRRI, Manila.

Kao, K.N. (1981) Plant formation from barley anther cultures with Ficoll media. Z. Pflanzenphysiol. 103: 437–443.

Kao, K.N. & M.R. Michayluk (1975) Nutritional requirements for growth of *Vicia hajastana* cells and protoplasts at a very low population density in liquid media. Planta 126: 105–110.

Kasha, K.J. (1986) Significance and prospects for haploidy in triticale and wheat. In: Proceedings of International Triticale Symposium, Sydney, 1986. Australian Institute of Agricultural Science, Occasional Publication No. 24: 9–21.

Keller, J. (1991) Influence of media conditioning on *in vitro* development in pollen suspensions of triticale. Biologisches Zentralblatt 110: 207–214.

Laurie, D.A. & M.D. Bennett (1988) The production of haploid wheat plants from wheat × maize crosses. Theor. Appl. Genet. 76: 393–397.

Lazar, M.D., G.W. Schaeffer & P.S. Baenziger (1984) Cultivar and cultivar × environment effects on the development of callus and polyhaploid plants from anther cultures of wheat. Theor. Appl. Genet. 67: 273–277.

Lehmann, C. (1989) Attempts to produce stabilized genotypes of tetraploid triticale (ABD)(ABD)RR by anther culture. In: XII EUCARPIA Congress 1989. Vorträge für Pflanzenzüchtg. 15: 7–11.

Linde-Laursen, I. (1973) Chromosome doubling of wheat × rye hybrids. Paper presented at the EUCARPIA Cereal Section Meeting on "Breeding of Triticale", Leningrad, 2–6 July, 1973.

Lukjanjuk, S.F. & S.A. Ignatova (1986) Triticale: Production of haploid and homozygous plants. In: Y.P.S. Bajaj (Ed.), Biotechnology in Agriculture and Forestry, Vol. 2. Crops I, pp. 532–543. Springer Verlag, Berlin.

Marsolais, A.A., G. Seguin-Swartz & K.J. Kasha (1984) The influence of anther cold pretreatments and donor plant genotypes on *in vitro* androgenesis in wheat. Plant Cell Tiss. Organ Cult. 3: 69–79.

Miller, T.E. (1987) Systematics and evolution. In: F.G.H. Lupton (Ed.), Wheat Breeding – Its Scientific Basis, pp. 1–30. Chapman and Hall, London/New York.

Murashige, T. & F. Skoog (1962) A revised medium for rapid growth and bioassays with tobacco tissue cultures. Physiol. Plant. 15: 473–497.

O'Donoughue, L.S. & M.D. Bennett (1994) Comparative responses of tetraploid wheats pollinated with *Zea mays* L. and *Hordeum bulbosum* L. Theor. Appl. Genet. 87: 673–680.

Ohkawa, Y., K. Suenaga & T. Ogawa (1992) Production of haploid wheat plants through pollination of sorghum pollen. Jpn. J. Breed. 42: 891–894.

Ouyang, J.W., S.M. Zhou & S.E. Jia (1983) The response of anther culture to culture temperature in *Triticum aestivum*. Theor. Appl. Genet. 66: 101–109.

Ryan, S.A., P.J. Larkin & F.W. Ellison (1987) Somaclonal variation in some agronomic and quality characters in wheat. Theor. Appl. Genet. 74: 77–82.

Ryöppy, P.H. (1991) Storage proteins of doubled haploid triticale lines. Vortr. Pflanzenzüchtg. 20: 163–167.

Ryöppy-Ekbom, P. & P.M.A. Tigerstedt (1984) Production of homozygous lines of triticale by anther culture. In: W. Lange, A.C. Zeven & N.G. Hogenboom (Eds.), Efficiency in Plant Breeding: Proceedings of the 10th Congress of the European Association for Research on Plant Breeding EUCARPIA, Wageningen, 19–24 June 1983, p. 345. Pudoc, Wageningen.

Ryöppy, P., J. Honkanen & P.M.A. Tigerstedt (1986a) Increasing the efficiency of triticale anther culture. In: Genetic Manipulation in Plant Breeding, pp. 339–341. Walter de Gruyter & Co., Berlin.

Ryöppy, P., P.M.A. Tigerstedt & J. Honkanen (1986b) Pretreatment experiments on triticale anther culture. In: Proceedings of International Triticale Symposium, Sydney 1986. Australian Institute of Agricultural Science, Occasional Publication No. 24: 434.

Snape, J.W., E. Simpson, M.D. Bennett & V. Chapman (1980) Haploid production in wheat by hybridization with *Hordeum bulbosum*. In: D.R. Davies & D.A. Hopwood (Eds.), The Plant Genome, p. 247. The John Innes Charity, Norwich.

Subrahmanyam, N.C. & K.J. Kasha (1975) Chromosome doubling of barley haploids by nitrous oxide and colchicine treatments. Can. J. Genet. Cytol. 17: 573–583.

Sun J.S., Z.Q. Zhu, J.J. Wang & P.M.A. Tigerstedt (1980) Studies on the anther culture of triticale. Acta Bot. Sin. 22: 27–31.

Sunderland, N. (1974) Anther culture as a means of haploid production. In: K.J. Kasha (Ed.), Haploids in Higher Plants: Advances and Potential. Proceedings of the First International Symposium, June 10–14, 1974, pp. 91–122. University of Guelph, Guelph.

Thomas, J.B. & P.J. Kaltsikes (1977) The effect of colchicine on chromosome pairing. Can. J. Genet. Cytol. 19: 231–249

Wang, C.C., C.S. Sun, C.C. Chu & S.C. Wu (1978) Studies on the albino pollen plantlets of rice. In: Proceedings of Symposium on Plant Tissue Culture, May 25–30, 1978, Peking, pp. 149–160. Science Press, Peking.

Wang, G., J. Ji, Y.-B. Wang, H. Hu, I.P. King & J.W. Snape (1993) The genetic characterization of novel multi-addition doubled haploid lines derived from triticale × wheat hybrids. Theor. Appl. Genet. 87: 531–536.

Wang Y.Y., C.S. Sun, C.C. Wang & N.F. Chien (1973) The induction of pollen plantlets of triticale and *Capsicum annuum* from anther culture. Sci. Sin. 16: 147–151.

Wang, X. & H. Hu (1985) The chromosome constitution of plants derived from pollen of hexaploid triticale × common wheat $F_1$ hybrids. Theor. Appl. Genet. 70: 92–96.

Wilson, A.S. (1876) On wheat and rye hybrids. Trans. Proc. Bot. Soc. Edinburgh 12: 286–288.

# 6. Haploidy in ryegrass

SVEN BODE ANDERSEN, STEN MADSEN, NIELS ROULUND, NIELS HALBERG and ANNETTE OLESEN

**Contents**



## 1. Introduction

Both *Lolium perenne* and *L. multiflorum* are strong outbreeders that exhibit severe inbreeding depression (Bean & Yok-Hwa, 1972; Matzk, 1974) and possess a strict gametophytic self-incompatibility system which involves two major gene loci with modification by the genetic background (Lawrence *et al.*, 1983).

The objectives of ryegrass breeding depend on their use either as forage or turf. The main breeding objective for forage *Lolium* types is to increase the dry matter yield. However, the quality of the dry matter, like voluntary intake and digestibility, have a considerable impact on animal production and must also be considered (Crampton *et al.*, 1960; Harrison *et al.*, 1984). For turf grasses a different spectrum of goals like swath density, breadth of leaf, persistence and leaf colour are important.

The basic breeding strategy in ryegrass has mainly involved various types of mass selection, followed by identification of superior clones with good combining ability for subsequent development of synthetic varieties. These methods, in which the selection is done in highly heterozygous plant material, have made considerable progress for some characters such as persistence and disease resistance. However, improvement in yield and other quantitatively inherited traits has been moderate (Sjödin, 1991; Van Wijk & Reheul, 1991). The objective of inbreeding of ryegrass has been to obtain more homozygous material (Eckmeier & Wricke, 1994) for the purpose of producing hybrid cultivars. When acceptable inbred lines of ryegrass can be produced, it may be possible to develop hybrid cultivars through exploitation of either

*S.M. Jain, S.K. Sopory & R.E. Veilleux (eds.),* In Vitro *Haploid Production in Higher Plants, Vol. 4,* 133–147.

cytoplasmic male sterility (Connolly & Wright-Turner, 1984) or self-incompatibility (England, 1974; Hayward *et al.*, 1991).

## 2. Methods for haploid induction in ryegrass

Reports on haploidy in *Lolium* have been rare until 1980. Spontaneous parthenogenetic haploids together with triploids, tetraploids, and pentaploids were found among plants from germinating seeds with twin embryos (Sulinovski *et al.*, 1983). Such haploids are generally found only in low frequencies and their identification has been laborious. Induction of seeds with parthenogenetic embryos via pollination with alien sources of pollen has not been reported in ryegrass. These techniques have been successful in barley (Jensen, 1977), wheat (Inagaki & Tahir, 1990), and potato (Uijtewaal *et al.*, 1987). It is possible that parthenogenetic haploids may be induced in *Lolium* through *in vitro* culture of unfertilized ovules. This method has been successful for several important cereal species like rice, barley, wheat and maize (Bajaj, 1990), but regeneration of viable haploid plants in *Lolium* through ovule/ovary culture has not so far been reported.

Attempts to produce haploids from cultured microspores in ryegrass was started by Clapham (1971) who reported regeneration of albino plants from Italian ryegrass. Until 1987, the results obtained with this and other *Lolium* species were poor thereby preventing systematic adaptation of anther culture (Table 1). Using methods and media adapted from wheat anther culture, Olesen *et al.* (1988) reported results from a screening experiment with 30 different genotypes of perennial ryegrass (*Lolium perenne* L.) They achieved

*Table 1.* Early results and screening experiments with anther cultures in *Lolium* ssp.

| Species | Response | Reference |
|---|---|---|
| *L. multiflorum* | albino plants | Clapham, 1971 |
| | green plants | Niizeki, 1977 |
| | green plants | Nitzsche & Wenzel, 1977 |
| | green plants | Pagniez & Demarly, 1979 |
| | green plants | Boppenmeier et al., 1989 |
| | green plants | Bante et al., 1990 |
| *L. temulentum* | albino plants | Rose et al., 1987a,b |
| *L. perenne* | albino plants | Stanis et al., 1983 |
| | green plants | Stanis & Butenko, 1984 |
| | green plants | Olesen et al., 1988 |
| | green plants | Boppenmeier et al., 1989 |
| | green plants | Bante et al., 1990 |

an average of 19.2 plants per 100 cultured anthers. Although the majority of plants produced was albino, three diploid ryegrass clones with a capacity to produce 1–3 green plants per 100 cultured anthers were identified. Shortly afterwards, similar results for green plant production from cultured anthers of *L. perenne* and *L. multiflorum* using other substrates and modified methods (Bante *et al.*, 1990) or modifications of the techniques used for barley anther culture (Boppenmeier *et al.*, 1989) were reported. General response level, however, even with the best ryegrass clones remained rather low (1–2 green plants/100 anthers) which made the optimization of media and protocol difficult.

## 3. Microspore-derived haploids in *Lolium*

A complete anther culture response in cereals and grasses (green plants/anther) can be described as a product of three different types of basic responses: a) embryo formation (embryos/anther), b) regeneration frequency (plants/embryo), and c) the frequency of green plants among regenerants. Media composition, culture methods and the genes of the plant material have different effects on embryo formation, regeneration and percentage of green plants. Their effects on anther response thus often become complex. An additional problem for the study of anther response is that, in many experiments, the low number of green plants actually obtained does not permit statistically verifiable conclusions. However, effects of media and methods on the components of anther culture can often be studied.

The main problem with large-scale production of microspore-derived plants in ryegrass is a tendency to regenerate albino plants. With recently developed anther culture methods in *Lolium*, embryos and plants can be obtained in large numbers from most genotypes; however, routine regeneration of green plants is restricted to certain clones. Formation of albino plants, which will not survive beyond *in vitro* culture, is a serious problem in anther cultures of most cereals and grasses. Albino cereal plants have been demonstrated to possess immature plastids, apparently blocked at various stages of development (Clapham, 1973; Sun *et al.*, 1974). Hybridization studies using DNA-probes have shown deletions or alterations in part of the chloroplast genome of albino wheat and barley plants (Day & Ellis, 1985). The capacity of some of the widely used model varieties like "Ciano" in wheat and "Igri" in barley to produce primarily green plants has been shown by reciprocal crosses to be under genetic control (Tuvesson *et al.*, 1989; Larsen *et al.*, 1991). Apart from these facts, nothing more is known about the nature of the albino problem.

## 4. Genotypic effects in *Lolium* anther culture

Genotypic effects on anther response is one of the best described features in anther culture of cereals. In particular, the capacity to form microspore-derived embryos from cultured anthers, and the frequency of green regenerants are genetically controlled (Andersen *et al.*, 1987; Knudsen *et al.*, 1989). Several ryegrass clones (Table 1) have similar genetic control of anther culture response because routine regeneration of green plants has been generally restricted to elite clones (Olesen *et al.*, 1988; Boppenmeier *et al.*, 1989; Bante *et al.*, 1990). Halberg *et al.* (1990) demonstrated clearly that some genes strongly affect anther culture in ryegrass. Ryegrass clones derived from crosses between three slightly responsive parents capable of producing 0.4–1.1 green plants per 100 cultured anthers showed strong transgressive segregation. Among 55 hybrid clones tested for anther culture response, 6 were superior to their parents. These superior clones produced from 11 to 59 green plants/100 cultured anthers, an improvement in anther culture response by 10–50 times the performance of the parents. Genetic improvement in regeneration capacity was less pronounced. These results strongly suggested that the original parental plants carried different genes affecting embryo formation and green plant formation. Crossing resulted in recombination of these genes and some of the hybrids showed higher anther culture response with improved capacity for both embryo and green plant formation. These results indicated a possible way to produce haploids in low responding ryegrass via the so called "inducer" approach. The idea is to cross anther culture responsive clones with non-responsive breeding material in order to obtain hybrid offspring with high anther culture response. If anther culture response in ryegrass shows some degree of genetic dominance, this approach will allow large scale production of microspore-derived plants from breeding material. Table 2 illustrates the magnitude of difference in anther culture response observed with ordinary breeding clones of perennial ryegrass and recombined high responding "inducers".

A more recent study has shown success of an "inducer" approach in perennial ryegrass (Madsen *et al.*, 1994). Hybrid offspring from crosses between randomly selected breeding clones of perennial ryegrass and highly responsive "inducer" clones produced on an average 2.2 green plants per 100 anthers. Only 0.3 green plants per 100 anthers on average were obtained from the breeding clones. This study also showed that the capacity of green plant formation, the limiting step in ryegrass anther culture, is controlled by the genes of donor plants. There are good indications of specific combining ability in the study (Madsen *et al.*, 1994) that capacity for green plant formation is controlled by a few major genes in perennial ryegrass.

Recent developments in DNA-based genetic marker systems (RFLPs, RAPDs, microsatellites) have provided improved possibilities for the detailed study of both qualitatively and quantitatively inherited traits (Rafalski & Tingey, 1993). The protocol consists of generating a population for mapping

*Table 2.* Anther response of 3 breeding clones and 3 highly responsive "inducer" clones of *Lolium perenne*. Average of two replicates with a total of 400 anthers cultured per genotype

| | Breeding clones | | | High responding inducer clones | | |
|---|---|---|---|---|---|---|
| | D7 | D21 | D24 | I3 | I8 | I10 |
| Embryos per 100 anthers | 262 | 4.0 | 208 | 66 | 63 | 165 |
| Plants regenerated per 100 embryos | 35.6 | 36.8 | 44.0 | 87.8 | 47.1 | 65.5 |
| Percentage of green plants | 0.0 | 0.0 | 1.1 | 31.9 | 51.7 | 47.7 |
| Green plants per 100 anthers | 0.0 | 0.0 | 1.3 | 18.6 | 15.7 | 47.4 |

($F_2$, backcross, or DH-lines) from a hybrid between parents with different genes for anther culture response. The segregating offspring are evaluated carefully for anther culture response and, subsequently, for segregation of marker genes. For each marker locus investigated, all offspring are divided into one group with one allele of the marker locus and another group with the alternate allele. If the two groups show significantly different anther culture response then the marker may be linked to a gene (QTL) affecting the trait. Using RFLP markers, Cowen *et al.* (1992) identified four different locations on the maize chromosomes that affected number of embryos and plants obtained from anther culture in one hybrid combination. Devaux & Zivy (1994) used 28 protein markers to study anther culture response from 50 DH-lines in barley and identified 4 chromosome locations which affected either embryo formation or regeneration/percent green plants. Andersen *et al.* (unpublished) used 50 DH-lines of barley selected particularly for the study of genes controlling green plant formation in barley anther culture in combination with around 100 RAPD markers. They were successful in dissecting genetic control of albinism into two major genes each linked to one RAPD marker in the barley cross studied. It is apparent from these studies that mapping populations of around 50 DH-lines and the study of 30–100 segregating genetic markers will efficiently detect the most important genes for anther culture response in a cross combination. Results from similar studies in ryegrass, particularly with respect to genes controlling capacity to form androgenic green plants, could become an efficient tool to produce haploids in ryegrass with the "inducer" approach in the future. If genetic markers linked to genes controlling green plant formation can be identified, they may be used to identify responsive clones among the offspring. This would greatly facilitate haploid production in ryegrass because resources can then be devoted to responsive material.

## 5. Methods and media for *Lolium* anther culture

Although exploitation of genes controlling anther culture response in ryegrass seems to have potential, it nevertheless restricts the choice of plant material for use in DH-techniques. Thus, improvements of media and methods to make ryegrass more anther culture responsive are highly desirable. Methods for the induction of androgenic haploids in *Lolium* have, mostly, been adapted from modified techniques developed for barley and wheat. Donor plants have been grown under controlled conditions in either glasshouse or growth chamber. Spikes containing microspores at the late uninucleate stage of development are selected for culture. This suitable stage of pollen development, generally, coincides with a particular morphological stage when half to three-quarters of the spike has emerged from the sheath. The determination of pollen developmental stage in ryegrass can be done with a modified protocol from wheat (He & Ouyang, 1984). Spikes excised from glasshouse or growth chamber-grown plants are surface sterilized for 8–10 min in 3% Korsolin (Bode Chemie GmbH & Co, Hamburg) or 0.1% mercuric chloride followed by three rinses in sterile water. Following 30–40 days of anther culture on an induction medium amended with 6–9% maltose or sucrose, 1.5 mg/l 2,4–D, and 0.5 mg/l kinetin, microspore-derived embryos emerge from responding anthers. For regeneration, embryos are transferred to a medium supplemented with low carbohydrate (3% sucrose) and growth regulators (0.25 mg/l of both NAA and kinetin) and allowed to regenerate plants for another 30-40 days. Green regenerants readily form roots on the regeneration medium but are often grown for another month on a medium devoid of growth regulators before transfer to soil under shaded glasshouse conditions.

Harvested spikes can be kept at 3–7°C for 3–4 days before anther culture without detectable reduction in anther response. Bante *et al.* (1990) reported that prolonged cold pretreatment of spikes before anther culture of *L. perenne* and *L. multiflorum* was beneficial in some genotypes. Culture temperature used for the induction of *Lolium* androgenic embryos is similar to that for barley or wheat. Results from an experiment with 26°C culture temperature and a heat pretreatment during the first 3 days of *Lolium* anther culture indicated that 3 day heat pretreatment routinely used for wheat anther culture (Ouyang *et al.*, 1983) was ineffective in both genotypes studied (Table 3). Both genotypes produced more embryos at 26°C compared with 22°C. However, clone 245 produced more green regenerants at the lower temperature. A constant culture temperature is routinely used for the induction of embryos in our laboratory.

*Table 3.* Embryo response and plant formation in anther culture of two genotypes of *Lolium perenne* grown at 33°C for 3 days followed by constant 26°C compared to constant 26°C during the entire culture period. After Olesen (1987)

| Clone | Culture temperature (°C) | No. of anthers cultured | Per 100 anthers cultured | | |
|---|---|---|---|---|---|
| | | | embryos | green plants | total plants |
| 245 | 33 3d→26 | 1763 | 108 | 0.5 | 40.4 |
| | 26 | 1863 | 286 | 3.1 | 83.4 |
| | 22 | 1925 | 230 | 4.9 | 58.9 |
| 255 | 33 3d→26 | 992 | 170 | 3.4 | 31.4 |
| | 26 | 1073 | 255 | 6.5 | 43.1 |
| | 22 | 1019 | 179 | 3.2 | 30.2 |

The most successful induction media for anther culture of perennial ryegrass are modifications of 190–2 medium often used for wheat anther culture (Halberg *et al.*, 1990) or FW medium widely used for barley anther culture (Boppenmeier *et al.*, 1989). Concentration and type of sugar influence most anther cultures. The replacement of sucrose with maltose has been the key to anther culture success in barley (Hunter, 1987). Also for ryegrass (Bante *et al.*, 1990) and wheat (Orchinsky *et al.*, 1990), maltose was superior to sucrose for androgenic embryo formation. Our study showed that *Lolium perenne* anthers produced more embryos on induction media augmented with maltose compared to sucrose (Table 4). The number of green regenerants per 100 anthers was more than doubled on the higher concentrations of maltose. For embryo formation, the optimal concentration of both sucrose and maltose was 9% in this experiment, while the regeneration capacity of

*Table 4.* Effect of concentration of maltose and sucrose on anther culture in *Lolium perenne*. Approximately 3000 anthers cultured per treatment

| | Embryos per 100 anthers | | Plants per 100 embryos | | Percent green plants | | Green plants per 100 anthers | | Albino plants per 100 anthers | |
|---|---|---|---|---|---|---|---|---|---|---|
| Conc | Malt | Suc | Malt | Suc | Malt | Suc | Malt | Suc | Malt | Suc |
| 3% | 123 | 69 | 17.5 | 20.9 | 23.3 | 22.9 | 7.0 | 6.9 | 16.9 | 12.4 |
| 6% | 161 | 124 | 23.3 | 18.9 | 42.0 | 32.8 | 16.6 | 8.5 | 22.7 | 15.3 |
| 9% | 182 | 154 | 21.7 | 13.4 | 45.9 | 46.0 | 21.2 | 8.3 | 18.8 | 12.9 |
| 12% | 91 | 16 | 17.9 | 8.4 | 52.5 | 32.6 | 17.6 | 2.6 | 12.4 | 4.0 |

*Table 5.* Effect of culture method and 2,4–D concentration on anther culture with a highly responding clone of *Lolium perenne*. Approximately 3000 anthers cultured for each treatment

| mg/l 2,4-D | Embryos per 100 anthers | Plants per 100 embryos | Percent green plants | Green plants/ 100 anthers |
|---|---|---|---|---|
| 0 | 471 | 25.2 | 30.7 | 41.3 |
| 0.375 | 639 | 38.3 | 46.2 | 124.9 |
| 0.750 | 708 | 44.0 | 49.8 | 148.8 |
| 1.500 | 605 | 44.9 | 60.9 | 172.4 |

the microspore-derived embryos was favoured in the presence of slightly lower sugar concentrations.

Plant growth regulators have probably been the most intensively investigated group of compounds in anther culture media. Although auxins and cytokinins are not mandatory for ryegrass anther culture, they are generally recognized as beneficial for the induction of embryos from cultured anthers. With the inclusion of different concentrations of 2,4–D in the 190–2 induction medium (Table 5) particularly, an improvement in the percentage of green regenerants was observed at the higher auxin concentrations. The highest concentration of 2,4–D (1.5 mg/l) produced the highest number (172.4) of green plants per 100 cultured anthers.

## 6. Isolated microspore culture

Recently, we have succeeded in embryo formation and plant regeneration from isolated microspore cultures of *Lolium perenne*. Although this technique is still in its infancy for the regeneration of green plants in large numbers, it still has potential for improvement because many combinations of media and methods can be accurately compared in small culture volumes (i.e., in multiwell plates). In *L. perenne*, microspores are readily isolated directly from intact spikelets with the use of a blender avoiding the tedious work of manual isolation of anthers. *Lolium* species are highly outbreeding and produce a large amount of pollen; thus, the spikelets from one spike of perennial ryegrass produce enough isolated microspores (about 150 000) for a 2 ml culture suspension. Table 6 shows the results from an experiment studying the effect of different concentrations of 2,4–D and PAA (phenyl acetic acid) on isolated microspore cultures of perennial ryegrass. The results show a clear stimulating effect of 0.38–0.75 mg/l 2,4–D on embryo and

*Table 6.* Effects of PAA and 2,4–D on isolated microspore culture in *Lolium perenne*. Total number of embryos and plants from 5 different 0.3 ml microspore cultures of each treatment. Green plants regenerated in brackets

| 2,4-D | | PAA 0 mg/l | PAA 1 mg/l | PAA 5 mg/l | PAA 10 mg/l | Mean |
|---|---|---|---|---|---|---|
| 0 mg/l | Embryos | 78 | 64 | 56 | 30 | 57 |
| | Plants | 0 | 3 | 1 | 1 | 1 |
| 0.375 mg/l | Embryos | 163 | 105 | 86 | 36 | 98 |
| | Plants | 4 | 2 | 1 | 2 | 2 |
| 0.75 mg/l | Embryos | 127 | 99 | 83 | 35 | 86 |
| | Plants | 1 | 1(1) | 3 | 3 | 2 |
| 1.5 mg/l | Embryos | 130 | 67 | 64 | 24 | 71 |
| | Plants | 5 | 2(1) | 2(1) | 1(1) | 3 |
| Means | Embryos | 124.5 | 83.75 | 72.25 | 31.25 | 78 |
| | Plants | 2.5 | 2.0 | 1.75 | 1.75 | 2 |

plantlet formation. In contrast, there is clear evidence of a negative effect on microspore culture by adding PAA to the culture medium.

## 7. Characteristics of microspore-derived plants of *Lolium*

### 7.1. *Ploidy and chromosome doubling*

Cytological analyses have shown that 60–80% of diploid *L. perenne* and *L. multiflorum* androgenic regenerants have been derived from diploid parents. However, the remaining plants were haploid with a low percentage of polyploids (Olesen *et al.*, 1988; Bante *et al.*, 1990; Hayward *et al.*, 1990). Boppenmaier *et al.* (1989) reported a considerably lower frequency (29%) of diploids. Plants derived from anther culture of tetraploid ryegrass consisted of about half diploids and half tetraploids, and some plants with higher ploidy (Bante *et al.*, 1990).

Several studies using isozyme markers have documented that the majority of anther-derived plants with parental ploidy level have arisen through spontaneous chromosome doubling during development. When anthers were cultured from diploid parent plants showing heterozygosity in one or more isozyme loci, then the regenerated plants carried only one of the parental isozyme alleles (Olesen *et al.*, 1988; Hayward *et al.*, 1990). In an investigation of the isozyme patterns of 913 plants derived from anther culture of diploid *L. perenne* donor plants that exhibited heterozygosity in one or more of

*Table 7.* Distorted segregation of alleles in GOT/2 and GOT/3 isozyme loci among green (G) and albino (A) plants derived by anther culture from diploid parent clones of *Lolium perenne* heterozygous at the loci. Chi$^2$-value is for 1:1 segreation of alleles

| GOT/3 | | | | | | |
|---|---|---|---|---|---|---|
| Clone | 175x255 7 | | 255x245 14 | | 255x175 6 | |
| allele | G | A | G | A | G | A |
| bb | 0 | 0 | 24 | 28 | 3 | 21 |
| cc | 47 | 83 | 7 | 13 | 13 | 45 |
| Chi$^2$ | 47.0 | 83.0 | 9.3 | 5.5 | 6.3 | 8.7 |
| | *** | *** | ** | * | * | ** |
| GOT/2 | | | | | | |
| Clone | 255x245 14 | | 175x255 9 | | 255x175 6 | |
| allele | G | A | G | A | G | A |
| aa | 2 | 3 | | 34 | 3 | 8 |
| bb | 29 | 35 | | 59 | 13 | 45 |
| Chi$^2$ | 23.5 | 26.9 | | 6.7 | 6.3 | 25.8 |
| | *** | *** | | ** | * | *** |

*,**,***: Significant at 5, 1, and 0.1 % level respectively

three unlinked isozyme loci (GPI/2, GOT/2, GOT/3), we found only three heterozygous plants. These heterozygotes showed the heterozygous banding pattern from their parents in both GPI/2 and GOT/2 and likely resulted by regeneration of unreduced microspores. High spontaneous rates of chromosome doubling in anther cultures of *Lolium* can be considered an advantage, since it minimizes the problem of chromosome doubling of haploids which is necessary for breeding purposes.

### 7.2. *Gametic selection*

A feature of ryegrass anther culture is that genetic markers often show allelic segregation that differs from the expected 1:1 Mendelian segregation ratio. We studied segregation of isozyme alleles among androgenic plants of diploid *L. perenne* and observed highly significant deviations from 1:1 ratios in 10 of 18 cases (Table 7). Such deviations from the expected 1:1 ratio could be the result of lethal genes or genes affecting the developmental rate during anther culture. Lethal or sublethal mutations at a locus linked to an isozyme marker produce distorted segregation in the marker locus because individuals carrying an unfavourable allele have low survival rates. On the contrary, favourable alleles in gene loci affecting *in vitro* culturability linked to a genetic marker producing distorted segregation because their host individual has increased survival rate. The most important difference between the two types of segregation distortion is that selection against lethal or sublethal alleles during androgenic haploid production will eliminate the inbreeding depression that occurs during traditional inbreeding. Selection for *in vitro* culturability, however, would not be expected to occur during traditional breeding programmes, and such selection may affect the plant material if the selected genes are agronomically undesirable. Hayward *et al.* (1990) have presented a good example of either lethal or major genes affecting anther culture response linked to isozyme markers. The relative importance of these two mechanisms causing distorted segregation is not yet known for ryegrass.

### 7.3. *Growth vigor and fertility*

Self-fertility of DH-clones derived from *Lolium* anther culture is expected to be low because such plants are homozygous for both major loci controlling self-incompatibility. This was found by Madsen *et al.* (1993) at the time of seed set by selfing in several generations from such *L. perenne* DH-clones. When such homozygous plants are selfed for several generations, the incompatibility system in *Lolium* may break down. This may be caused by a strong selection on the stigmas of rare pollen grains carrying mutations favouring self-compatibility.

Generally, the majority of haploid microspore-derived as well as spontaneously chromosome doubled ryegrass plants are weak and slow growing compared to the parental material. Opsahl-Ferstad (1993) evaluated 140 DH-clones derived from three different parental clones together with their parents in a replicated field trial. The total forage yield of DH-clones was only 20–21% of their parents. DH-clones were more erect and sensitive to stem rust than their parents. In addition, an average seed yield of the DH-material was 8–12% of the parental lines. Such strong inbreeding depression is expected when completely homozygous DH-plants are produced in an outcrossing species like *Lolium*. Indeed many of the green regenerants from anther cultures of perennial ryegrass are weak and unable to survive until maturity. Some of the regenerants grow vigorously and can be multiplied for field evaluation. We cloned 75 such vigorous DH-plants and their parents, and evaluated them for cross fertility in a topcross design with five different open-pollinated ryegrass cultivars (as pollen donors). In spite of the vigorous growth of this material, most of the DH-clones showed poor seed set. Only 15 of the DH-clones (20%) produced enough seed to evaluate the topcross offspring in the subsequent field trials. On average, they produced 0.62 g seed/plant compared with 2.14 g/parental plant, and only one DH-clone produced more seed than parents. This exceptional clone produced 5.08 g seed/plant compared with 1.44 g/plant obtained from its parent clone, and was by far the most fertile genotype in these experimental trials. Topcross offspring from the 15 fertile DH-clones and their parents are currently being evaluated for yield in field trials.

## 8. Haploidy and breeding of ryegrass

Our experience with completely homozygous DH-clones of ryegrass suggests two types of inbreeding depression in this material: a) that affecting vegetative growth, and b) that affecting fertility. The majority of inbred genotypes can be eliminated by discarding DH-clones with poor vegetative growth. Often, the selected vigorously growing clones suffer an additional inbreeding depression leading to suppression of seed set to such an extent that they become useless for breeding purposes. Ironically, the poor performance of the majority of these homozygous ryegrass genotypes indicates their potential in future breeding of this species. The selection of rare DH-clones with exceptional growth, vigour, fertility, and disease resistance may lead to the elimination of agronomically undesirable alleles in the plant material. Such clones in the future may be used for the construction of narrow based synthetic or hybrid cultivars with improved quality and yield.

## 9. Conclusion

So far, haploid induction in ryegrass has been done with anther and isolated microspore cultures. Currently, the major hurdle in succeeful *Lolium* anther culture is the formation of albino plants. Most breeding materials demonstrate less than 1% of regenerated green plants. The capacity to form high frequencies of both embryos and green plants in *Lolium* anther culture is clearly affected by genes in the plant material. Such genes can be recombined to develop highly responsive types capable of regenerating 30–60 green plants per 100 anthers. High responding clones may subsequently be used as "inducers" for crosses with ordinary breeding material because the hybrid offspring on average respond with about 2 green plants per 100 cultured anthers.

Media and methods for anther culture in *Lolium* are similar to those used for barley and wheat with minor modifications of temperature, sugar type and concentration, and growth regulators. Isolated microspore culture has been achieved in *Lolium perenne*. Generally, 60–80% of the plants derived from anther culture of *Lolium* have spontaneously doubled their chromosome number and are completely homozygous. Such plants display severe inbreeding depression, as expected of DH-clones from a strong outbreeder in terms of both vegetative growth and fertility. Exceptional homozygous genotypes with good vegetative growth and high seed yield, however, can be identified. The potential of such exceptional genotypes with low inbreeding depression for future ryegrass breeding is under investigation.

In the future, for increasing *Lolium* haploid induction, efforts should be made to improve both the culture media and anther response, and to identify genetic markers for controlling green plant formation.

## 10 Acknowledgements

The authors are grateful to Miss Heidi Fårup, Miss Britta Henriksen, and Mr. Benny Kastrup for technical assistance. Financial support from the Danish Ministry of Agriculture, The Directorate for Agricultural Development is warmly acknowledged.

## 11. References

Andersen, S.B., I.K. Due & A. Olesen (1987) The response of anther culture in a genetically wide material of winter wheat (*Triticum aestivum* L.). Plant Breed. 99: 181–186.

Bante I., T. Sonke, R.F. Tandler, A.M.R. van den Bruel & E.M. Meyer (1990) Anther culture of *Lolium perenne* and *Lolium multiflorum*. In: R.S. Sangwan and B.S. Sangwan-Norreel (Eds.), The Impact of Biotechnology in Agriculture, pp. 105–127. Kluwer Academic Publishers, Dordrecht.

Bajaj, Y.P.S. (1990) *In vitro* production of haploids and their use in cell genetics and plant breeding. In: Y.P.S. Bajaj (Ed.), Biotechnology in Agriculture and Forestry 12. Haploids in Crop Improvement I, pp. 3–44. Springer-Verlag, Berlin.

Bean, E.W. & C. Yok-Hwa (1972) An analysis of the growth of inbred progeny of *Lolium*. J. Agric. Sci. Camb. 79: 147–153.

Boppenmeier, J., S. Züchner & B. Foroughi-Wehr (1989) Haploid production from barley yellow dwarf virus resistant clones of *Lolium*. Plant Breed. 103: 216–220.

Clapham, D. (1971) *In vitro* development of callus from the pollen of *Lolium* and *Hordeum*. Z. Pflanzenzüchtg. 65: 285–292.

Clapham, D. (1973) Haploid *Hordeum* plants from anthers *in vitro*. Z. Pflanzenzüchtg. 69: 142–155.

Connolly, W. & R. Wright-Turner (1984) Induction of cytoplasmic male-sterility into ryegrass (*Lolium perenne*). Theor. Appl. Genet. 68: 449–453.

Cowen, N.M., C.D. Johnson, K. Armstrong, M. Miller, A. Woosley, S. Pescitelli, M. Skokut, S. Belmar & J.F. Petolino (1992) Mapping genes conditioning *in vitro* androgenesis in maize using RFLP analysis. Theor. Appl. Genet. 84: 720–724.

Crampton, E.W., E. Donefer & L.E. Lloyd (1960) A nutritive value index for forage. J. Anim. Sci. 19: 538–544.

Day, A. & T.H.N. Ellis (1985) Deleted forms of plastid DNA in albino plants from cereal anther culture. Curr. Genet. 9: 671–678.

Devaux, P. & M. Zivy (1994) Protein markers for anther culturability in barley. Theor. Appl. Genet. 88: 701–706.

Eckmeier, F. & G. Wricke (1994) Hybridzüchtung mit Hilfe der genetischen Selbstinkompatibilität bei *Lolium multiflorum*. Vortr. Pflanzenzüchtg. 28: 286–288.

England, F.W.J. (1974) The use of incompatibility for the production of $F_1$-hybrids in forage grasses. Heredity 32: 183–188.

Halberg, N., A. Olesen, I.K.D. Tuvesson & S.B. Andersen (1990) Genotypes of perennial ryegrass (*Lolium perenne* L.) with high anther-culture response through hybridization. Plant Breed. 105: 89–94.

Harrison, A.M., A. Pike, P.D. Putwain & T. McNeilly (1984) Variation in digestibility in nine

populations of perennial ryegrass (*Lolium perenne*) and its genetic basis. Euphytica 33: 497–506.
Hayward, M.D., A. Olesen, I.K. Due, R. Jenkins & P. Morris (1990) Segregation of isozyme marker loci amongst androgenetic plants of *Lolium perenne* L. Plant Breed. 104: 68–71.
Hayward, M.D., D. Thorogood & J.G. Jones (1991) Manipulation and exploitation of the reproductive system in forage breeding. In: A.P.M. den Nijs & A. Algersma (Eds.), Fodder Crops Breeding: Achievements, Novel Strategies and Biotechnology, pp. 89–94. Pudoc, Wageningen.
He, D.G. & J.W. Ouyang (1984) Callus and plant formation from cultured wheat anthers at different developmental stages. Plant Sci. Lett. 33: 71–79.
Hunter, C.P. (1987) Plant generation method. European Patent Application, Nr. 0 245 898 A2, 1–8.
Inagaki, M. & M. Tahir (1990) Comparison of haploid production frequencies in wheat varieties crossed with *Hordeum bulbosum* L. and maize. Jpn. J. Breed. 40: 209–216.
Jensen, C.J. (1977) Monoploid production by chromosome elimination. In: J. Reinert & Y.P.S. Bajaj (Eds.), Applied and Fundamental Aspects of Plant Cell, Tissue, and Organ Culture, pp. 299–330. Springer-Verlag, Berlin.
Knudsen, S., I.K. Due & S.B. Andersen (1989) Components of response in barley anther culture. Plant Breed. 103: 241–246.
Larsen, E.T., I.K.D. Tuvesson & S.B. Andersen (1991) Nuclear genes affecting percentage of green plants in barley (*Hordeum vulgare* L.) anther culture. Theor. Appl. Genet. 82: 417–420.
Lawrence, M.J., C.H. Fearon, M.A. Cornish & M.D. Hayward (1983) The genetical control of self-incompatibility in ryegrasses. Heredity 51: 461–466.
Madsen S., A. Olesen & S.B. Andersen (1993) Self-fertile doubled haploid plants of perennial ryegrass (*Lolium perenne* L.). Plant Breed. 110: 323–327.
Madsen, S., A. Olesen, B. Dennis & S.B. Andersen (1995) Inheritance of anther culture response in perennial ryegrass (*Lolium perenne* L.). Plant Breed. 114: 165–168.
Matzk, F. (1974) Selbstfertilität, Inzuchtdepression und Kreuzungstechnische Experimente bei *Lolium*. Arch. Züchtungsforsch. 4: 141–151.
Niizeki, M. (1977) Haploid, polyploid and aneuploid plants from cultured anthers and calluses in species of *Nicotiana* and forage crops. J. Fac. Agric. Hokkaido Univ. 58: 343–466.
Nitzsche, W. & G. Wenzel (1977) Haploids in plant breeding. Fortschritte der Pflanzenzüchtung 8: 46–48. Advances in Plant Breeding. Verlag Paul Parey, Berlin.
Olesen, A. (1987) Anther culture of perennial ryegrass (*Lolium perenne* L.): The production of haploid plants and their potential in relation to traditional breeding strategies. Ph.D. Thesis, The Royal Veterinary and Agricultural University, Copenhagen.
Olesen, A., S.B. Andersen & I.K. Due (1988) Anther culture response in perennial ryegrass (*Lolium perenne* L.). Plant Breed. 101: 60–65.
Opsahl-Ferstad, H.G. (1993) Androgenetic response in grasses. I. Anther culture in perennial ryegrass (*Lolium perenne* L.). II. Molecular studies of embryogenesis in barley (*Hordeum vulgare* L.). Ph.D. Thesis, Agricultural University of Norway, Dept. of Biotechnological Sciences.
Orchinsky, B.R., L. McGregor, G.I.E. Johnson, P. Hucl & K.K. Kartha (1990) Improved embryoid induction and green shoot regeneration from wheat anthers cultured in medium with maltose. Plant Cell Rep. 9: 365–369.
Ouyang, J.W., S.M. Zhou & S.E. Jia (1983) The response of anther culture to culture temperature in *Triticum aestivum*. Theor. Appl. Genet. 66: 101–109.
Pagniez, M. & Y. Demarly (1979) Obtention d'individus androgénètiques par culture *in vitro* d'anthères de ray-grass d'Italie (*Lolium multiflorum* Lam.). Ann. Amélior. Plantes 29: 631–637.
Rafalski, J.A. & S.V. Tingey (1993) Genetic diagnostics in plant breeding: RAPDs, microsatellites, and machines. Trends Genet. 9(8): 275–280.
Rose, J.B., J.M. Dunwell & N. Sunderland (1987a) Anther culture of *Lolium temulentum*,

*Festuca pratensis* and *Lolium* × *Festuca* hybrids. I. Influence of pretreatment, culture medium and culture incubation conditions on callus production and differentiation. Ann. Bot. 60: 121–201.

Rose, J.B., J.M. Dunwell & N. Sunderland (1987b) Anther culture of *Lolium temulentum*, *Festuca pratensis* and *Lolium* × *Festuca* hybrids. II. Anther and pollen development *in vivo* and *in vitro*. Ann. Bot. 60: 203–214.

Sjödin J. (1991) Achievements in fodder crops breeding in nordic Europe. In: A.P.M. den Nijs & A. Algersma (Eds.), Fodder Crops Breeding: Achievements, Novel Strategies and Biotechnology, pp. 7–12. Pudoc, Wageningen.

Stanis, V.A. & R.G. Butenko (1984) Developing viable haploid plants in anther culture of ryegrass. Doklady Biol. Sci. 275: 249–251.

Stanis, V.A., A.K., Sliesaravichus & V.A. Kaleda (1983) [Regeneration of ryegrass and clover plants in tissue culture of different origin] (in Russian). Sel'Skokhozyaistvennaya Biologiya 5: 29–31.

Sulinovski, S., H. Wisniewska & K. Sekowska (1983) Frequency of spontaneous polyploids in *Lolium perenne* and *Festuca pratensis*. (Bibliographic citation): The Utilization of Genetic Resources in Fodder Crop Breeding, 13–16 September 1982, Aberystwyth, UK, 1983, pp. 55–59, 2 ref. Welsh Plant Breeding Station, Aberystwyth, UK.

Sun, C.S., C.C. Wang & C.C. Chu (1974) The ultrastructure of plastids in the albino pollen-plants of rice. Sci. Sin. 17: 794–797.

Tuvesson, I.K.D., S. Pedersen & S.B. Andersen (1989) Nuclear genes affecting albinism in wheat (*Triticum aestivum* L.) anther culture. Theor. Appl. Genet. 78: 879–883.

Uijtewaal, B.A., D.J. Huigen & J.G.T Hermsen (1987) Production of potato monohaploids ($2n = x = 12$) through prickle pollination. Theor. Appl. Genet. 73: 751–758.

Van Wijk, A.J.P. & D. Reheul (1991) Achievements in fodder crops breeding in maritime Europe. In: A.P.M. den Nijs & A. Algersma (Eds.), Fodder Crops Breeding: Achievements, Novel Strategies and Biotechnology, pp. 13–18. Pudoc, Wageningen.

# 7. Haploidy in sorghum

GEORGE H. LIANG, XU GU, GUILAN YUE, Z.S. SHI and K.D. KOFOID

**Contents**



## 1. Introduction

Haploid plantlets can be produced through various means (Clapham, 1977). However, haploid production using anther or microspore culture is amenable to many dicotyledonous and monocotyledonous crop species. The doubled haploid breeding scheme has advantages in accelerating and simplifying procedures for production of homozygous plants. Haploid cell suspension culture can also be used to screen for stress tolerance. In addition, callus, cell suspensions, or protoplasts derived from haploid cells can be used as recipients in transformation either by cocultivation with *Agrobacterium* or by a biolistic gun.

Nevertheless, haploid production in certain cereal crops, particularly *Sorghum*, remains difficult. This difficulty is due to low regeneration potential, especially after an extended period of culture; occurrence of albinos (Rose *et al.*, 1986); lack of response in callus induction from many elite cultivars and inbreds (Wen *et al.*, 1991); pigment exudation (phenolic compounds) from somatic cells (Cai & Butler, 1990; Guo & Liang, 1993); nonrandom generation of genotypes (Snape *et al.*, 1992); and cost factors. However, successful haploid production in sorghum, though rare, is known (Kumaravadivel & Rangasamy, 1994).

Few theoretical guidelines exist for tissue culture operations in general and many questions remain. Why are callus induction and plant regeneration easier in dicots, particularly the *Solanaceae*, than monocots? Why does anther culture work relatively well in rice, canola, and wheat but not in sorghum? Is anther culture or microspore culture under the control of rare but major genes or polygenes (De Buyser *et al.*, 1992)? Which cell (generative or vegetative) in the male gametophyte differentiates into a plantlet (Hassawi *et al.*, 1990)? What are the appropriate kinds and amounts of growth regulators for high-frequency callus induction and plant regeneration for each crop species? How do nutrient components interact with each other in the

*S.M. Jain, S.K. Sopory & R.E. Veilleux (eds.),* In Vitro *Haploid Production in Higher Plants, Vol. 4,* 149–161.

medium? Is sucrose, glucose, or maltose the best sugar for maintaining osmotic pressure and providing carbon to the growing cells? Should natural extracts be used in certain cases or only well-defined media? What are the most appropriate culture conditions (temperature, photoperiod, light intensity, and light quality) for callus induction and plantlet regeneration? How can the regenerated plantlets be transferred to soil with minimal loss? What causes the so-called somaclonal variation (Larkin & Scowcroft, 1981) and how can it be eliminated or reduced? How can all the variables involved in tissue culture be controlled to make the results more consistent and reproducible? What is the agronomic performance of doubled haploid cultivars or inbred lines derived from anther culture programs (Brown & Wernsman, 1982; Brown *et al.*, 1983; Campbell & Wernsman, 1994; Henry & De Buyser, 1990; Walker & Aycock, 1994)? Until some or all of these questions are answered and the variables controlled, the application of tissue culture techniques in practical operations will be limited. Haploid production in sorghum through anther or microspore culture remains a great challenge to many researchers.

Sorghum [*Sorghum bicolor* L. (Moench), subspecies *S. bicolor*, 2n = 2x = 20] is the fifth most important cereal grain crop in the world. The genus *Sorghum* is diverse and consists of five sections: *Eusorghum* or *Sorghum* (2n = 20, 40), *Parasorghum* (2n = 10), *Stiposorghum* (2n = 10), *Chaetosorghum* (2n = 40), and *Heterosorghum* (2n = 40). The haploid chromosome number in the genus *Sorghum* varies from 5 to 20. Compatible crosses between species across various sections, however, have never been reported. The phylogenetic relationships of the genus *Sorghum* have been analyzed from cytological and morphological data (Celarier, 1958; Garber, 1950; De Wet, 1978; Gu *et al.*, 1984; Wu, 1990) and also from recently developed molecular genetic techniques using chloroplast DNA (Duvall & Doebley, 1990), mitochondrial DNA (Guo *et al.*, 1996) and nuclear ribosomal DNA (Sun *et al.*, 1994).

Within the section *Eusorghum*, *S. bicolor* is believed to have originated in Africa (Mann *et al.*, 1983) and includes grain and forage sorghum types. Grain sorghum is one of the most important cereal crops in arid regions. It is used for food, beverages and vinegar, pulp, and animal feed. In China, the reddish brown pigments extracted from the glumes are used as dyes in foods and the stalks are processed for special building materials.

Due to the heterosis observed in $F_1$ hybrids, all commercial sorghums in the USA, China, Australia and many other countries are produced as $F_1$ hybrids using the $A_1$ male sterile cytoplasm and the corresponding nuclear fertility restorers. The pedigree system is the primary breeding method for inbred line development in sorghum. Population improvement and the conversion of exotic (tall and late) germplasm (Stephens *et al.*, 1967) provide base materials for further selection. To speed up the production of elite homozygous inbred lines, haploid production through the use of anther or microspore culture is desirable.

To produce haploids via tissue culture, the media applied to sorghum anther culture have been potato medium (Rose *et al.*, 1986), MS (Murashige & Skoog, 1962), or N6 (Chu *et al.*, 1975; Kumaravadivel & Rangasamy, 1994) media or modifications thereof with supplements of growth regulators and sugars. These media apparently are ineffective in inhibiting or neutralizing the exudation of pigments onto the media and the pigments, which are phenolic in nature, are detrimental to cells (Cai & Butler, 1990). For successful production of sorghum haploid plantlets using anther culture, a special component(s) that is able to prevent, bind or neutralize the pigments appears to be a prerequisite.

In this report, we intend to review the available information on sorghum haploid production through anther culture, the barriers encountered, the application of haploid research, and what needs to be done in the future.

## 2. General procedures

### 2.1. *Anther culture*

As for many other cereals, the appropriate time to place sorghum anthers on the nutrient medium is when microspores reach the mid- to late-uninucleate stage (Wen *et al.*, 1991; Kumaravadivel & Rangasamy, 1994). In sorghum, anthers in the upper portions of the panicle reach maturity first. A few anthers from upper florets should be sampled to examine microspore development and to determine the microspore stages under a microscope and then to establish correlations between the uninucleate stage and some morphological traits, such as the distance between flag leaf and the second leaf. Such a relationship should facilitate the anther culture procedure. Once the relationship between microspore development and a morphological trait is established for a genotype, further microscopic examination for microspore stage should no longer be necessary.

After the panicle branches with anthers containing microspores at the uninucleate stage are collected either from field- or greenhouse-grown plants, they are sterilized with 60% EtOH for 3 min in a laminar air flow hood (Wen *et al.*, 1991), or they can be sterilized for 15 min in a solution of 0.2% sodium hypochlorite (NaOCl) followed by rinses (Rose *et al.*, 1986), or rinsed with 70% ethanol and then sterilized in $HgCl_2$ for 1 min (Kumaravadivel & Rangasamy, 1994). Then the anthers are removed from each floret and placed on a nutrient medium in Petri dishes (100 × 15 mm). Each Petri dish usually contains about 30 to 40 anthers. Caution should be taken not to injure the anther sac. The Petri dishes are sealed with parafilm and kept in an incubation chamber in darkness at 28 to 30 ± 1°C until calli are formed, at which time calli are transferred to a regeneration medium.

Unlike rice, wheat, or barley, pretreatment of sorghum panicles at low temperatures (4–10°C) does not seem to be beneficial for callus induction

*Table 1.* Medium composition for callus induction in sorghum anther culture

| Components | Medium MS-t-z-2 | C-17-2 | MS-d-4 |
|---|---|---|---|
| Major elements (mg/l) | | | |
| $KNO_3$ | 1,900 | 1,400 | 1,900 |
| $KH_2PO_4$ | 170 | 400 | 170 |
| $NH_4NO_3$ | 1,650 | 300 | 1,650 |
| $CaCl_2$ $2H_2O$ | 440 | 150 | 440 |
| $MgSO_4$ $7H_2O$ | 370 | 150 | 300 |
| Minor elements (mg/l) | | | |
| $Na_2$-EDTA | 37.3 | 37.3 | 37.3 |
| $FeSO_4$ $7H_2O$ | 27.8 | 27.8 | 27.8 |
| $MnSO_4$ $H_2O$ | 17.1 | 8.6 | 17.1 |
| $ZnSO_4$ $7H_2O$ | 8.6 | 8.6 | 8.6 |
| $H_3BO_3$ | 6.2 | 6.2 | 6.2 |
| Trace elements (mg/l) | | | |
| KI | 0.83 | 0.83 | 0.83 |
| $CuSO_4$ $5H_2O$ | 0.025 | 0.025 | 0.025 |
| $CoCl_2$ $6H_2O$ | 0.05 | 0.025 | 0.025 |
| $NaMoO_4$ $2H_2O$ | 0.25 | -- | 0.25 |
| Vitamins (mg/l) | | | |
| Glycine | 2 | 2 | 2.0 |
| Thiamine HCl | 0.1 | 1 | 0.1 |
| Pyridoxine HCl | 0.5 | 0.5 | 0.5 |
| Nicotinic acid | 0.5 | 0.5 | 0.5 |
| Biotin | -- | 1.5 | -- |
| Myo-inositol | 100 | -- | 100 |
| Growth regulators (mg) | | | |
| 2,4-D | 3.0 | 2 | -- |
| Kinetin | 0.3 | 1 | 2.5 |
| Zeatin | 2.2 | -- | -- |
| IAA | -- | -- | 2.0 |
| Sucrose (g/l) | 20 | 20 | 30 |
| Difco agar (g/l) | 6.8 | 6.8 | 7.0 |
| pH | 5.8 | 5.8 | 5.8 |

Adapted from Wen *et al.*, 1991. *Euphytica* 52: 177–181.

(Rose *et al.*, 1986). On the contrary, high temperature (30°C) treatment has been reported to be effective for callus induction from anthers (Kumaravadivel & Rangasamy, 1994).

Two types of basic media have been used for callus induction in sorghum, the potato medium with 9% sucrose (Rose *et al.*, 1986) and chemically defined media MS-t-z-2, C-17–2 (Table 1) (Wen *et al.*, 1991) or $N_6$. The former has the advantages of simplicity and low cost, and the latter is amenable to further improvement by adjusting quantity and kinds of nutrient elements.

Among the auxins, 2,4–D has been used most frequently in the range of 9 to 13.5 μM, whereas NAA does not seem to have been as effective. Kinetin, in the range of 1.4 to 4.7 μM, has been the most often used cytokinin. For callus redifferentiation or plant regeneration, half strength

MS with 2% sucrose (MS 1/2) and stepwise transfer of calli to MS 1/2 media first containing 2,4–D (9 μM), then BA (5.2 μM) plus 2,4–D (4.5 μM), and finally 2,4–D (0.45 μM) were successful (Rose *et al.*, 1986). $B_5$ vitamins (Gamborg, 1975) may be added and IAA (1.7 μM) may replace 2,4–D (Kumaravadivel & Rangasamy, 1994). However, from more than 1,000 pieces of calli regenerated using these techniques, Rose *et al.* (1986) obtained only four albino plants and the chromosome number of these plants was not documented. On the other hand, MS medium supplemented with 11.4 μM IAA and 11.75 μM kinetin (MS-d-4) (Table 1) was reported to be an effective regeneration medium (Wen *et al.*, 1991). Nevertheless, all 29 regenerated plants were diploid with 10 bivalents (Fig. 1A) and highly fertile. Using modified MS medium and the hybrid, CSH5, Kumaravadivel & Rangasamy (1994) reported that both diploid and haploid plants had been regenerated through anther culture. However, no microscopic examination of the chromosome number, which would confirm the haploidy, was presented and the doubled haploids may have been the products of somatic tissues.

To enhance root development of regenerated seedlings, a liquid MS medium without sucrose and growth regulators has been used for regenerated seedlings in a growth chamber at 16°C constant temperature. A well-developed root system is essential to reduce the loss of regenerated seedlings when transferred. Temperature also appears to be a critical factor for successful transfer of seedlings from medium to pots containing vermiculite or to a culture solution. Use of a hydroponic solution has also facilitated the collection of root tips for chromosome counts (Wen *et al.*, 1991).

To identify the ploidy level of regenerated seedlings, root tips of seedlings derived from pollen calli should be examined for chromosome number. Root tips 1 cm in length can be excised from the seedlings and pretreated with saturated naphthalene monobromide solution for 2 h and then fixed for 2 days in 3 parts ethanol:1 part propionic acid (v/v) and 1% $FeCl_3$ (w/v). After the root tips are removed from the fixative, they are stained in acetocarmine for 1 h, rinsed in distilled water, and submerged in 4% (w/v) pectinase (Sigma) solution at room temperature (22°C). A volume of 0.5 ml pectinase solution can be used to treat four to five root tips. The enzyme treatment is terminated by replacing the enzyme solution with 45% acetic acid (Tang & Liang, 1987). Root tips are squashed on a slide and examined under a microscope. Results have been satisfactory (Fig. 1B).

Alternatively, the root tips can be treated in 1 N HCl or a 3 EtOH:1 acetic acid mixture at room temperature for 7 to 10 days before staining in acetocarmine and squashing on a slide (Yu & Liang, unpublished). This method is used when enzymes are not available, but lengthens the time needed to complete the chromosome counting process.

### 2.2. *Microspore culture*

In sorghum, microspore culture may provide a way to overcome pigment exudation from cells of anther walls or other somatic tissues and the regen-

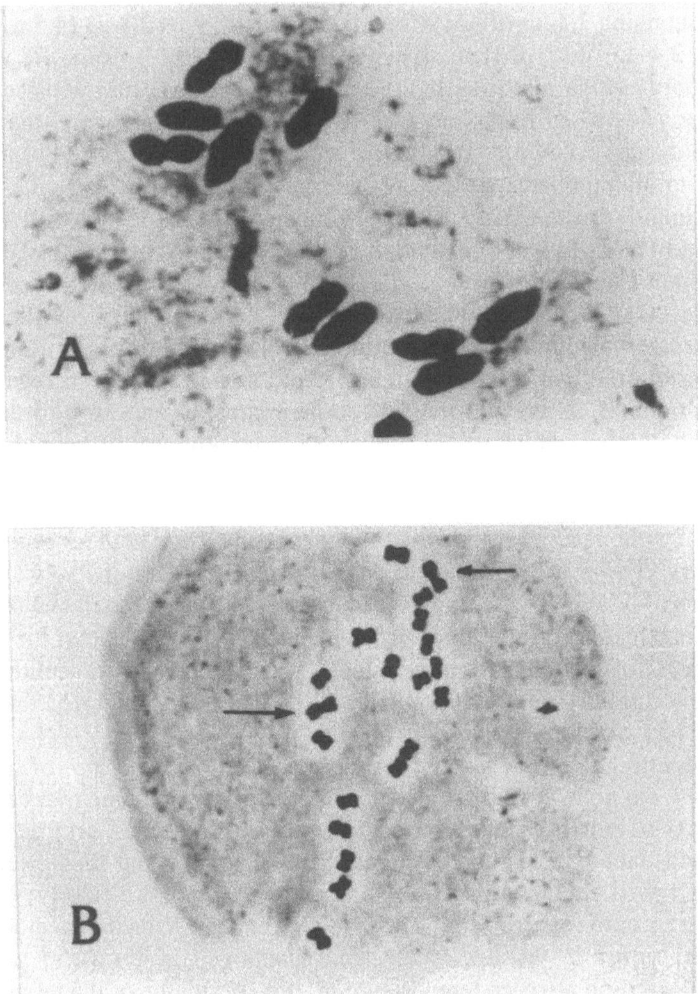


*Figure 1.* (A) Ten bivalents shown from the microsporocyte collected from anther culture regenerated sorghum plants, indicating the maternal origin of those plants. (B) A metaphase cell of *S. bicolor* showing 20 chromosomes; the longest chromosome pair is indicated by the arrows.

eration from cells of anther tissue and filaments that are diploid in nature. However, medium composition and culture operations are more complex for culture of microspores than for whole anthers. In addition, microspore culture experiments can encounter unpredictability, low repeatability, and low embryo yield, as in *Brassica* and wheat (Duijs *et al.*, 1992; Oberthur *et al.*, 1993).

Microspores can be extracted from anthers in three ways: 1) floating the anthers on liquid medium supplemented with Ficoll; collecting microspores after anther dehiscence by sieving and centrifugation; and resuspending the microspores in a given medium (Sunderland & Roberts, 1977; Kao *et al.*, 1991), 2) isolating the microspores mechanically by gentle maceration with a mortar and a pestle or a blender; sieving through a nylon mesh (pore size 52 μm); rinsing with a washing solution (0.3 M mannitol, 5 mM $CaCl_2$, 5 mM MES); collecting microspores by centrifugation (1,000 × g); and resuspending in a medium at a final density of $1.5 \times 10^4$ to $2 \times 10^5$ microspores per ml (Hoekstra *et al.*, 1992; Nichterlein & Friedt, 1993; Tuversson & Ohlund, 1993), and 3) culturing the anthers on a medium for 4 to 5 days, transfer those anthers to a liquid medium, shake using a vortex mixer and then collect the microspores by centrifuging and resuspending on an agar medium (Liang, unpublished). Likewise, the sessile florets can be placed directly into a Waring micro-blender and blended at high speed for 12 to 15 s. Then the blended slurry is filtered through a stainless steel mesh (325 μm) followed by a 52 μm nylon mesh. The filtrate is collected and centrifuged at 1,000 × g, the pellets are washed with 50 ml culture medium, and centrifuged again, resuspended in 3 ml culture medium, and plated on a solidified medium. However, positive experimental results from microspore culture of sorghum anthers are not yet available.

## 3. Discussion and prospects

Breeding schemes using haploid materials could be highly efficient because effects of both dominant and recessive genes can be observed and selected; predicted gains are high due to greater additive genetic variance in doubled haploid lines than in random mating lines; interspecific hybridization between ploidy-reduced (e.g., diploids from autotetraploids) plants and natural diploids can be made readily; and only one generation is required to obtain homozygosity. For plant transformation research using haploid cells, protoplasts, or calli, no segregation is expected among chromosome-doubled transgenic plants. Cytogenetically, haploids are sources of aneuploids, chromosome affinity in the single genome complement can be analyzed by observing bivalents, and chromosome pairing relationships and fertility of polyhaploids (haploids from polyploid species) and, hence, the nature of the ploidy (auto- versus allo- and segmental polyploids) can be evaluated. Cytologically, the existence, structure, and behavior of the synaptonemal complex in haploid cells can be examined and compared to that in diploid tissues. For developmental genetics, haploid cells, their multiplication, and their redifferentiation are useful for studies of embryo development and activities of homeobox genes in single doses (Ma *et al.*, 1994).

Clearly, different species require different media for successful haploid generation through anther culture. After repeated trials, those media designed for rice, wheat, and maize have not produced the same results in sorghum (Liang, unpublished). Apparently, sorghum needs a unique medium composition to neutralize the normal pathway of the microspores so that they will develop into calli or, more desirably, directly into haploid plantlets.

Among the medium components, one or more must be able to inhibit, bind, or prevent the cellular exudation of dark purple pigments, which contain large amounts of polyphenols, such as flavan-4–ols. Sorghum cells in culture are susceptible to polyphenols. Typically, sorghum kernels with brown or purple testa (or with purple plant color but no testa) produce purple pigments that diffuse into the medium shortly after inoculation. On the other hand, at least some immature embryos and kernels without brown or purple testa are known to produce callus without pigments (Guo & Liang, 1993). Likewise, plants with tan plant color, such as Tx 430, do not exudate dark purple pigments. However, many elite inbreds and high-yielding hybrids have kernels with brown and purple testa that contain tannins and can reduce bird damage because of their bitter taste and also make better liquor than those without colored testa.

We are testing compounds that may be able to neutralize the negative effects of polyphenols. One of the compounds, PVP 360 (Sigma), is known to bind phenolic compounds, and our initial results showed that PVP (1 mg/ml) indeed reduced the amount of dark pigments exuding from the anthers. The other compound, PTU (15 mg/l), is also being tested. Perhaps, effects of harmful polyphenols can be neutralized by one or more of the chemicals, and haploid calli can be produced from anther culture. On the other hand, microspore culture and wide crosses, where maternal haploids are likely to be induced, may provide a different approach to overcome the pigments' effect. However, the essential medium composition remains to be worked out before the pigments can be neutralized or bound by PVP.

Because the endogenous hormonal level and kinds of hormones involved during microspore formation could vary in different tissues or organs and from different species, various growth regulators at different concentrations should be tested for haploid generation in sorghum. In addition to 2,4–D and kinetin, CPA, dicamba, NAA, IAA, IBA, picloram, BA, and zeatin should be tested at various concentrations and in various combinations.

Alternatively, haploid plantlets may be produced through wide crosses, such as sorghum × maize crosses. Laurie & Bennett (1988) reported that haploids can be produced from wheat pollinated by maize. A mean of 21% embryo recovery with 81% plantlet regeneration was found (Mujeeb-Kazi *et al.*, 1991), and 17% haploids were obtained in a similar experiment with spring wheat (Rines *et al.*, 1991). The advantages of using maize pollen to produce haploids in wide crosses are the lesser effect of genotype dependency, a higher frequency of haploids, and few albinos. However, the flowering times of winter wheat and maize are different, and emasculation is time consuming. Therefore, unless maize pollen can be stored by cryopreservation for a long period of time and safe chemical hybridization agents or male gametocides are available, wide crosses using maize pollen still have limits. No reports are known of sorghum haploids produced with maize pollen.

Although sorghum is grown in temperate zones, it has a tropical origin (Mann *et al.*, 1983; Rangan & Vasil, 1984). It is an extremely variable crop, and anther, microspore, or embryo cultures are genotype-dependent (Ma *et al.*, 1987). This implies that, when different inbreds or hybrids are tested, the appropriate culture conditions may vary among genotypes. In general, tissue cultures for crop species of tropical origin are more successful at higher temperatures and medium length of photoperiod. The exact requirements for temperature and photoperiod regimes in sorghum anther culture should be tested systematically.

The cultivated form *Sorghum bicolor* (2n = 20) and its relative, *S. halepense* (L.) Pers. (2n = 40), a grassy wild sorghum, are intercrossable. Genomic analysis of their hybrids provides evidence that *S. bicolor* may be a tetraploid and *S. halepense* an octoploid (Tang & Liang, 1988). Frequency of chromosome pairing relationships and fertility levels of haploids from *S. bicolor* and *S. halepense* should provide useful information on the level of their ploidy.

Furthermore, is *S. halepense* an autopolyploid or a segmental polyploid? Genomic analysis suggests that it is a segmental polyploid, whereas tetrasomic inheritance (Casady & Anderson, 1952) and numerical analysis of chromosome pairing in interspecific hybrids between *S. bicolor* and *S. halepense* (Liang, unpublished) indicate that it may be an autopolyploid. Chromosome behavior in polyhaploids derived from *S. halpense* through anther culture could provide additional data on the nature of ploidy.

Somaclonal variation, or spontaneous genetic alteration occurring in cells growing *in vitro*, is well known in many crops (Larkin & Scowcroft, 1981; Lee & Phillips, 1988) but its mechanism is not clearly understood except that the variation is principally a result of either chromosome number variation, somatic meiosis (Wen *et al.*, 1989), deletion, or rearrangements. However, most of the variations are not beneficial to breeding programs because they already exist in natural populations.

The stability of organellar genomes of tissue culture regenerants could be analyzed molecularly using DNA restriction fragment length polymorphism (Chowdhury & Vasil, 1993). The presence in regenerated plants of subgenomic configurations that are undetectable in donor parents appears to be both time- and organ/tissue dependent (Marère-Le Paven *et al.*, 1992). Both chloroplast and mitochondrial DNA of regenerated plants can be analyzed, using DNA probes or fraction I protein as genetic markers (Kung, 1983).

Microspore culture can be a valuable atlernative to produce doubled haploid plants and be incorporated in breeding programs, transformation research and in embryo development analysis. In cereals, plant regeneration from isolated microspores is known for wheat (Mejza *et al.*, 1993) and barley (Hoekstra *et al.*, 1992). Microspore culture is also successful in *Brassica* species (Baillie *et al.* 1992; Duijs *et al.*, 1992; Hansen & Svinnset, 1993; Takahata *et al.*, 1993), linseed (Nichterlein and Friedt, 1993) and other crop species.

Using the barley cv. Igri, it was found that microspore culture is 5 times more efficient than anther culture (Hoekstra *et al.*, 1992). For wheat, microspores isolated from pretreated anthers (25°C for 2 days) of a range of genotypes and cocultured with wheat ovaries on a maltose containing medium produced fertile plants (Mejza *et al.*, 1993).

The final analysis of the merit for using anther and microspore culture in breeding programs would be the field evaluation of the cultivars developed from the tissue culture programs. There are some successful examples. Jinghua No. 1, a highly successful wheat cultivar selected from a 3-way cross [(Lovrin 18 × 5238–036) × Hongliang No. 4] using anther culture, was shorter than the parental lines and yielded 7.3 to 32.1% more than the standard cultivar Nongda 139 in multiple year and multiple location tests. On average, Jinghua No. 1 produced 5 tons/ha and was highly resistant to stripe rust, leaf rust, and powdery mildew, which are important diseases in northern China. Jinghua No. 1 represents a superior product from 800,000 microspores or 400 anthers placed on a medium in which 46 green plants were regenerated (Hu, 1986). Likewise, a winter wheat cultivar Florin was released in France using anther culture from $F_1$ plants of a cross between Wizard and Iena (Henry & De Buyser, 1990). Florin is a semidwarf, high yielding cultivar and more importantly, it saved at least 4 years in time (the initial cross was made in 1978 and Florin was registered in 1985) compared to conventional breeding schemes. In another experiments where 40 lines each derived by anther culture and by single-seed descent were compared in a spring wheat cross, an acceptable level of performance could be found among the anther-culture-derived lines, and selected lines were as agronomically desirable as those produced by single seed descent (Mitchell *et al.*, 1992).

Selection for disease resistance can also be facilitated by the haploid approach. In attempting to generate multiple-disease resistant dihaploids, screening for tobacco mosaic virus and wild fire resistance at haploid stage was found to be effective in eliminating the susceptible genotypes early (Walker & Aycock, 1994). Some dihaploids derived from anther culture performed at a high level for one or more agronomic traits and exhibited resistance to several diseases, suggesting that multiple resistance can be obtained through anther culture. Similarly, selection for resistance to black shank disease, caused by a soil borne pathogen, was reported to be more efficient in $F_1$-derived haploid population than selection within an $F_2$ diploid population (Campbell & Wernsman, 1994). The advantages of haploid selection are apparent when space becomes a limiting factor. For traits that can be assessed on vegetative structures, haploid selection offers promise in many crop species.

## 4. Conclusions

The significance and implications of haploid production and utilization are well-recognized. Progress has been made in many crops including some cereals. However, confirmed haploid plantlets have been rare through anther or microspore culture in sorghum. Clearly, the genus *Sorghum* has special requirements for *in vitro* culture. More importantly, why is sorghum so different from other cereals? Does sorghum represent a different evolutionary line, or has a major mutation deterred the microspores from forming calli for regeneration? Many questions remain to be answered. Therefore, a great challenge is facing us – the somatic cell geneticists, the plant breeders, the physiologists, the botanists, the evolutionists, and the molecular biologists – to produce haploids successfully in sorghum for genetic, cytogenetic, and breeding operations.

## 5. References


Baillie, A.M.R., D.J. Epp, D. Hutcheson & W.A. Keller (1992) *In vitro* culture of isolated microspores and regeneration of plants in *Brassica campestris*. Plant Cell Rep. 11: 234–237.

Brown, J.S., E.A. Wernsman & R.J. Schnell, II (1983) Effect of a second cycle of anther culture on flue-cured lines of tobacco. Crop Sci. 23: 729–733.

Brown, J.S. & E.A. Wernsman (1982) Nature of reduced productivity of anther-derived dihaploid lines of flue-cured tobacco. Crop Sci. 22: 1–5.

Cai, T.S. & L. Butler (1990) Plant regeneration from embryogenic callus initiated from immature inflorescences of several high-tannin sorghum. Plant Cell Tiss. Organ Cult. 20: 101–110.

Campbell, K.G. & E.A. Wernsman (1994) Selection among haploid sporophytes for resistance to black shank in tobacco. Crop Sci. 34: 662–667.

Casady, A.J. & K.L. Anderson (1952) Hybridization, cytological and inheritance studies of a sorghum cross – autotetraploid sudangrass × (Johnsongrass × 4*n* sudangrass). Agron. J. 44: 189–194.

Celarier, R.P. (1958) Cytotaxonomy of the Andropogoneae III. Subtribe Sorgeae, Genus *Sorghum*. Bull. Torrey Bot. Club 85: 49–62.

Chowdhury, M.K.U. & I.K. Vasil (1993) Molecular analysis of plants regenerated from embryogenic cultures of hybrid sugarcane cultivars (*Saccharum* spp.). Theor. Appl. Genet. 86: 181–188.

Chu, C.C., C.C. Wang, C.S. Sun, C. Hsu, K.C. Yin, C.Y. Chu & F.Y. Bi (1975) Establishment of an efficient medium for anther culture of rice through comparative experiments on the nitrogen sources. Sci. Sin. 18: 659–668.

Clapham, D.H. (1977) Haploid production in cereals. In: J. Reinert & Y.P.S. Bajaj (Eds.), Plant Cell, Tissue, and Organ Culture, pp. 279–298. Springer-Verlag, New York.

De Buyser, J., S. Hachemi-Rachedi, M.L. Lemee, S. Sejourne, J.L. Marcotte & Y. Henry (1992) Aneuploid analysis of anther culture response in wheat. Plant Breed. 109: 339–342.

De Wet, J.M.J. (1978) Systematics and evolution of *Sorghum* Sect. *Sorghum* (Gramineae). Am. J. Bot. 65: 477–484.

Duijs, J.G., R.E. Voorrips, D.L. Visser & J.M.B. Custers (1992) Microspore culture is successful in most crop types of *Brassica oleracea* L. Euphytica 60: 45–55.

Duvall, M.R. & J.F. Doebley (1990) Restriction site variation in the chloroplast genome of *Sorghum* (Poaceae). Syst. Bot. 15: 472–80.

Gamborg, O.L. (1975) Callus and cell culture, Append. II. In: O.L. Gamborg & L.R. Wetter (Eds.), Plant Tissue Culture Methods, p. 91. Nat'l Res Council Canada, Saskatoon, Sask.

Garber, E.D. (1950) Cytotaxonomic studies in the genus *Sorghum*. Univ. Calif. Pub. Bull. 23: 283–361.

Gu, M.H., H.T. Ma & G.H. Liang (1984) Karyotype analysis of seven species in the genus *Sorghum*. J. Hered. 75: 196–202.

Guo, J.H., D.Z. Skinner & G.H. Liang. Philogenetic relationships among sorghum taxa inferred from mitochondrial DNA restriction fragment analysis. Genome (in press).

Guo, J.H. & George H. Liang (1993) Callus induction and plant regeneration of cultivated and wild sorghums. Cytologia 58: 203–210.

Hansen, M. & K. Svinnset (1993) Microspore culture of swede (*Brassica napus* ssp. *rapifera*) and the effects of fresh and conditioned media. Plant Cell Rep. 12: 496–500.

Hassawi, D., R.G. Sears & G.H. Liang (1990) Microspore development in the anther culture of wheat. Cytologia 55: 475–478.

Henry, Y. & J. De Buyser (1990) Wheat anther culture: Agronomic performance of doubled haploid lines and the release of a new variety "Florin". In: Y.P.S. Bajaj (Ed.), Biotechnology in Agriculture and Forestry 13: Wheat, pp. 285–352. Springer-Verlag, New York.

Hoekstra, S., M.H. van Zijderveld, J.D. Louwerse, F. Heidekamp & F. van der Mark (1992) Anther and microspore culture of *Hordeum vulgare* L. cv. Igri. Plant Sci. 86: 89–96.

Hu, Daofen (1986) Jinghua No. 1, a winter wheat variety derived from pollen sporophyte. In: H. Hu & H. Yang (Eds.), Haploidy of Higher Plants *in Vitro*, pp. 136–148. China Acad. Publ., Beijing/Springer-Verlag, New York.

Kao, K.N., M. Saleen, S. Abrams, M. Pedras, D. Horn & C. Mallard (1991) Culture conditions for induction of green plants from barley microspores by anther culture methods. Plant Cell Rep. 9: 595–601.

Kumaravadivel, N. & S.R.S. Rangasamy (1994) Plant regeneration from sorghum anther cultures and field evaluation of progeny. Plant Cell Rep. 13: 286–290.

Kung, S.D. (1983) Fraction-1 protein and chloroplast DNA as genetic markers. In: D.A. Evans, W.R. Sharp, P.V. Ammirato & Y. Yamada (Eds.), Handbook of Plant Cell Culture I, pp. 583–606. Macmillan Publishing Co., New York.

Larkin, P.J. & W.R. Scowcroft (1981) Somaclonal variation – a novel source of variability from cell cultures for plant improvement. Theor. Appl. Genet. 60: 197–214.

Laurie, D.A. & M.D. Bennett (1988) The production of haploid wheat plants from wheat × maize crosses. Theor. Appl. Genet. 76: 393–397.

Lee, M. & R.L. Phillips (1988) The chromosomal basis of somaclonal variation. Ann. Rev. Plant Physiol. Plant Mol. Biol. 39: 413–437.

Ma, H., M. Gu & G.H. Liang (1987) Plant regeneration from cultured immature embryos of *Sorghum bicolor* (L.) Moench. Theor. Appl. Genet. 73: 389–394.

Ma, H.C., M.D. McMullen & J.J. Finer (1994) Identification of a homeobox-containing gene with enhanced expression during soybean (*Glycine max* L.) somatic embryo development. Plant Mol. Biol. 24: 465–473.

Mann, J.A., C.T. Kimber & F.R. Miller (1983) The origin and early cultivation of sorghums in Africa. Tx. Agr. Exp. Sta. Bull. No. 1454.

Marère-Le Paven, M.C., J. De Buyser, Y. Henry, F. Corre, C. Hartmann & A. Rode (1992) Multiple patterns of mtDNA reorganization in plants regenerated from different *in vitro* cultured explants of a single wheat variety. Theor. Appl. Genet. 85: 9–14.

Mejza, S.J., V. Morgant, D.E. DiBona & J.R. Wong (1993) Plant regeneration from isolated microspores of *Triticum aestivum*. Plant Cell Rep. 12: 149–153.

Mitchell, J.M., R.H. Busch & H.W. Rines (1992) Comparison of lines derived by anther culture and single-seed descent in a spring wheat cross. Crop Sci. 32: 1446–1451.

Mujeeb-Kazi, A., O. Riera-Lizarazu, R. Delgado & S. Cano (1991) Diverse applications of maize mediated polyhaploid production in the Triticeae. Agron. Abstr., p. 107. American Society of Agronomy, Madison, WI.

Murashige T. & F. Skoog (1962) A revised medium for rapid growth and bioassays with tobacco tissue cultures. Physiol. Plant. 15: 437–497.

Nichterlein, K. & W. Friedt (1993) Plant regeneration from isolated microspores of linseed (*Linum usitatissimum* L.). Plant Cell Rep. 12: 426–430.

Oberthur, L.E., P.S. Baenziger, D.M. Wesenberg & S. Kachman (1993) The inheritance of gametoclonal variation in spring wheat doubled haploid derived lines. Agron. Abstr., p. 96. American Society of Agronomy, Madison, WI.

Rangan, T.S. & I.K. Vasil (1984) Millets. In: D.A. Evans, W.R. Sharp, P.V. Ammirato & Y. Yamada (Eds.), Handbook of Plant Cell Culture III, pp. 126–150. Macmillan Publishing Co., New York.

Rines, H.W., D.W. Davis & R.H. Busch (1991) Use of male steriles in producing haploid wheat and oat plants by widecross hybridizations. Agron. Abstr., p. 114. American Society of Agronomy, Madison, WI.

Rose, J.B., J.M. Dunwell & N. Sunderland (1986) Anther culture of *Sorghum bicolor* (L.) Moench. I. Effects of panicle pretreatment, anther incubation, temperature, and 2,4-D concentration. Plant Cell Tiss. Organ Cult. 6: 23–31.

Snape, J.W., J.W. Ouyang, B.B. Parker & S.E. Jia (1992) Evidence for genotypic selection in wheat during the development of recombinant inbred lines by anther culture and single seed descent. J. Genet. Breed. 46: 167–172.

Stephens, J.C., F.R. Miller & D.T. Rosenow (1967) Conversion of alien early combine genotypes. Crop Sci. 7: 396.

Sun, Y., D.Z. Skinner, G.H. Liang & S.H. Hulbert (1994) Phylogenetic analysis of *Sorghum* and related taxa using internal transcribed spacers of nuclear ribosomal DNA. Theor. Appl. Genet. 89: 26–32.

Sunderland, N. & M. Roberts (1977) New approach to pollen culture. Nature 270: 236–238.

Takahata, Y., Y. Takani & N. Kaizuma (1993) Determination of microspore population to obtain high frequency embryogenesis in broccoli (*Brassica oleracea* L.). Plant Tissue Cult. Lett. 10: 49–53.

Tang, H. & G.H. Liang (1987) An improved technique for cytological observations and occurrence of polysomaticism in sorghum root tips. J. Hered. 78: 51–53.

Tang, H. & G.H. Liang (1988) The genomic relationship between cultivated sorghum [*Sorghum bicolor* (L.) Moench] and Johnson grass [*S. halepense* (L.) Pers.]. Theor. Appl. Genet. 76: 277–284.

Tuversson, I.K.D. & R.C.V. Ohlund (1993) Plant regeneration through culture of isolated microspores of *Triticum aestivum* L. Plant Cell Tiss. Organ Cult. 34: 163–167.

Walker, D.R. & M.K. Aycock, Jr. (1994) Development of anther-derived dihaploids to combine disease resistance in Maryland tobacco. Crop Sci. 34: 335–338.

Wen, F., F.L. Barnett & G.H. Liang (1989) Somatic pairing and separation of chromosomes in root cells of regenerated sorghum plants. J. Hered. 80: 159–160.

Wen, F.S., E.L. Sorensen, F.L. Barnett & G.H. Liang (1991) Callus induction and plant regeneration from anther and inflorescence culture of *Sorghum*. Euphytica 52: 177–181.

Wu, T.P. (1990) *Sorghum macrospermum* and its relationship to the cultivated species *S. bicolor*. Cytologia 55: 141–151.

# 8. Haploid induction in buckwheat (*Fagopyrum esculentum* Moench)

BORUT BOHANEC

**Contents**



## 1. Introduction

Buckwheat (*Fagopyrum esculentum* Moench) is a dicotyledonous plant that produces a grain used in a similar way as bread cereals; however, its taste, color and other distinctive features make buckwheat unique. Buckwheat is an undemanding crop with a short growing period and can be sown as a "catch" crop, after harvesting the main crop, or as a main crop in regions with a short growing season. Nutritional tests have shown that buckwheat has the highest biological value of its protein (over 90%) among plants (Eggum *et al.*, 1980). This is due to the well-balanced amino acid composition of protein, and especially high lysine content (Javornik *et al.*, 1981). It was traditionally a food of poor people and has thus retained importance as a starch and protein source in many of the less developed countries. Today it is recognised as a very important element in the modern diet also in developed countries.

Buckwheat is an open-pollinated crop with strong self-incompatibility. Flowers exhibit heterostylism under control of a super gene: thrum *Ss* and pin *ss*. Self-pollination has only rarely been achieved – especially difficult in thrum plants. Present buckwheat cultivars are based on the sib-mating technique and usually express a high degree of heterozygosity resulting in phenotypic diversity within a cultivar. The use of dihaploid plants would therefore open new possibilities in buckwheat breeding with regard to the homogeneity of lines and possible expression of vigour in hybrids. We can also speculate that use of doubled haploids would help to elucidate the buckwheat genetic constitution, especially heterostily mechanisms, since doubled haploid thrum plants would be "superthrum" possessing homozygous dominant *SS* super-

*S.M. Jain, S.K. Sopory & R.E. Veilleux (eds.),* In Vitro *Haploid Production in Higher Plants, Vol. 4,* 163–170.

gene. This potential was mentioned by Nešković *et al.* (1986). There are not many research reports on buckwheat anther culture and, as far as we know, there is no report on culture of unpollinated ovules. Adachi *et al.* (1988), and recently Fanchu *et al.* (1992) reported the possibility of regeneration from anthers but in both cases the haploid nature of regenerants was not confirmed. During the last few years we have been studying methods for both androgenic and gynogenic regeneration of buckwheat. In this paper we briefly summarize our results with androgenesis and present some results with unpollinated ovule culture.

## 2. Haploid induction by anther culture

Anther culture experiments performed in our laboratory for several years were mainly focused on determining the optimal medium for stable induction and regeneration of haploid plants. Experiments were performed with diploid cultivars, mainly 'Darja', though, occasionally 'Siva' and 'Rana 60' were used. Details on the protocols were published by Bohanec *et al.* (1993). Culture procedures included testing of many different induction media and different physical conditions. On the basis of culturing over 20.000 anthers the procedure described below was found optimal for androgenic regeneration of buckwheat plants.

### 2.1. *Inoculation technique*

Donor plants are preferably grown in the field, when day temperatures are lower than 20°C, so spring or autumn seasons are suitable. When grown in a greenhouse, additional illumination (850 $\mu$mol $m^{-2}s^{-1}$) should be used. Inflorescences are taken when pollen is in the uninucleate or early binucleate stage; this stage usually corresponds to 0.3–0.5 mm length anthers. Inflorescences are sterilised for 10 min in 5% NaOCl with addition of a few drops of Tween 20. Since flower buds within an inflorescence differ in size, only anthers from appropriate flowers can be inoculated. Usually it is possible to use 15–20 flower buds from 1 plant. Anthers are placed in 35 mm Petri dishes and kept in the dark at 25 ± 1°C until regeneration.

### 2.2. *Selected media*

Optimal induction and regeneration media according to our studies have the composition presented in Table 1. Regeneration usually occurs on induction medium (A) with 30 days and regenerants are transferred to a medium B to avoid unnecessary hyperhidricity. Further subculturing is done on medium C and rooting is achieved on medium D.

*Table 1.* Composition of media used for the induction and regeneration of haploids using anther culture

| | A | B | C | D |
|---|---|---|---|---|
| Murashige & Skoog 1962 macro and micro elements | + | + | + | 1/2 strength |
| m-inositol | 100 mg $l^{-1}$ | 100 mg $l^{-1}$ | 100 mg $l^{-1}$ | 100 mg $l^{-1}$ |
| thiamine | 2 mg $l^{-1}$ | 2 mg $l^{-1}$ | 2 mg $l^{-1}$ | 2 mg $l^{-1}$ |
| maltose | 90 g $l^{-1}$ | - | - | - |
| sucrose | - | 30 g $l^{-1}$ | 30 g $l^{-1}$ | 15 g $l^{-1}$ |
| coconut milk | - | 15% | - | - |
| benzyladenine (BA) | 11.10 μM | 4.44 μM | 4.44 μM | - |
| indole-3-acetic acid (IAA) | 2.85 μM | 1.14 μM | 1.14 μM | - |
| indole-3-butyric acid (IBA) | - | - | - | 2.46 μM |
| gellan-gum | 2 g $l^{-1}$ | - | - | - |
| agar | - | 8 g $l^{-1}$ | 8 mg $l^{-1}$ | 8 mg $l^{-1}$ |
| pH | 5.7 | 5.7 | 5.7 | 5.7 |

A = medium for induction and regeneration of buds/embryos from anthers, B = medium for establishment of shoot culture, C = medium for subculture of shoots, D = medium for root induction

## 2.3. *Observations of anther development*

Following inoculation, buckwheat anthers respond in three different ways; one part remains unchanged, one part elongates and one part swells. Further development was observed only from swollen anthers. Regeneration usually occurred as bud formation from anthers, and that was accompanied by callus development to a certain extent. Induction of embryoid formation directly from anthers was very rare.

## 2.4. *Induction frequency*

Generally the induction frequency was low. Several experiments resulted only in callus formation without regeneration. In the most successful experiment, induction frequency was 0.5% on previously described media and growing conditions. We conclude, that at this stage it is possible to obtain a limited number of regenerants from anther culture but further experiments are needed to improve induction frequency.

Ploidy level of regenerants was determined by counting the chromosomes in the root tips of *in vitro* rooted regenerants. Only some regenerants (8) exhibited the haploid chromosome number, the rest being diploid, tetraploid or mixoploid (22). Microscopic observations of developing microspores during culture revealed some unusual nuclear divisions, such as formation of two vegetative nuclei or formation of two equal uninucleate cells.

## 3. Haploid induction by ovule culture

Ovule culture experiments for haploid induction were initiated to determine the possibility of inducing a higher percentage of regenerants via gynogenesis than via androgenesis. Based on results with *Nicotiana* (Kumashiro & Oinuma, 1985; Wernsman, 1992) we also speculated that gynogenic regenerants would have less genetic aberrations than androgenic plants.

### 3.1. *Materials and methods*

The main cultivar tested was the diploid Rana 60. Donor plants were grown in either the field or the greenhouse with additional illumination (Philips SON T lamps 850 μmol $m^{-2}s^{-1}$). Flowers were collected just before anthesis or immediately after anthesis. They were sterilised for 10 min in 5% NaOCl; both heterostylic forms (pin and thrum) were included in the experiments and were inoculated separately. Five ovules per Petri dish (35 mm, containing 5 ml of culture medium) were cultured on selected media in a growth chamber at 25 ± 1°C in the dark until regeneration.

Basal medium was that of Gamborg *et al.* (1968) $B_5$ macro and micro elements, and 100 mg $l^{-1}$ inositol, 2 mg $l^{-1}$ thiamine, 1 mg $l^{-1}$ pyridoxine, 1 mg $l^{-1}$ nicotinic acid, 2 mg $l^{-1}$ glycine, 2.85 μM IAA, 2 g $l^{-1}$ gellan-gum. Growth regulator and carbohydrate concentrations varied as indicated in

*Table 2.* Regeneration from ovules inoculated on different induction media – no. of ovules cultured, number of calluses induced, no. of embryos or shoots regenerated, no. of regenerants that survived as stable shoots in culture

| μM | g $l^{-1}$ | no. of ovules | no. of calluses induced | no. of regenerants | no. of surviving shoots |
|---|---|---|---|---|---|
| 2iP 24.60, IAA 2.85, | maltose 90 | 317 | 3 (0.95%) | 1 (0.32%) | 1 (0.32%) |
| 2iP 24.60, IAA 2.85, | sucrose 85.5 | 312 | 0 | 0 | 0 |
| 2iP 24.60, 2,4-D 9.05, | maltose 90 | 335 | 1 (0.30%) | 0 | 0 |
| 2iP 24.60, 2,4-D 9.05, | sucrose 85.5 | 340 | 5 (1.47%) | 0 | 0 |

Part B: Effect of different concentrations of two cytokinins (BA and 2iP) on regeneration

| μM | μM | no. of ovules | no. of calluses induced | no. of regenerants | no. of surviving shoots |
|---|---|---|---|---|---|
| 2iP 24.60 (coautoclaved), | IAA 2.85 | 255 | 0 | 0 | 0 |
| 2iP 24.60, | IAA 2.85 | 614 | 5 (0.81%) | 11 (1.79%) | 4 (0.65%) |
| 2iP 49.20, | IAA 2.85 | 354 | 7 (1.98%) | 12 (3.39%) | 5 (1.41%) |
| BA 5.55, | IAA 2.85 | 375 | 17 (4.53%) | 2 (0.53%) | 1 (0.27%) |
| BA 11.10, | IAA 2.85 | 1.391 | 27 (1.94%) | 22 (1.58%) | 10 (0.72%) |
| BA 22.19, | IAA 2.85 | 370 | 2 (0.54%) | 8 (2.16%) | 3 (0.81%) |

Table 2. Media were sterilised by autoclaving; 6–($\gamma$,$\gamma$-dimethylallylamino) purine (2iP) was either filter sterilised and added to the medium after autoclaving or, coautoclaved as indicated; pH was adjusted to 6.0 before autoclaving.

After inoculation of ovules the following data were recorded: number of callus inductions, number of embryo or shoot regenerations, number of regenerants that survived until the normal shoot culture stage.

Further procedures, including post regeneration treatment and acclimatization were similar to those of Bohanec & Neškoviæ, 1991. Chromosomal analysis was done as described by Sinković & Bohanec, 1988. For flow cytometry analysis of intact *in vitro* grown tissues, leaves were cut (Galbraith *et al.*, 1983) and stained with propidium iodide (Reed & Wernsman, 1989).

### 3.2. *Results*

The preliminary test was performed by culturing 1200 ovules on basal medium supplemented with 24.6 $\mu$M 2iP (coautoclaved) and 90 g $l^{-1}$ maltose. The observation (data not shown) was, that most ovules after initial enlargement gradually changed from white to brown in the next 3–4 weeks. At this stage no further changes occurred, except for regeneration of two embryos that developed shoots on subculture (Fig. 1). Further experiments were conducted to test the effects of different auxins, cytokinins and carbohydrate sources (Table 2).

### 3.3. *Ploidy level of regenerants*

The ploidy level was determined by chromosome counts in root tips developed *in vitro*. Roots of haploid plants were weak, making chromosome analysis difficult. The presence of haploid cells ($2n = 1x = 32$) was confirmed in 10 of 24 regenerated genotypes. A strong tendency toward endoreduplication was noticed. Ploidy level was analysed again by flow cytometry on leaves of plantlets grown *in vitro* after approximately 18 months. The result was that 13 regenerants were diploid ($2n = 2x = 16$) and 4 were tetraploid ($2n = 4x = 32$) (Fig. 2). One possible explanation was that ploidy level was restored from haploid to diploid or tetraploid level upon further *in vitro* culture.

### 3.4. *Discussion*

Our results demonstrate that regeneration can be obtained from culture of unpollinated ovules of buckwheat. We have concluded that maltose is superior to sucrose at equimolar concentrations, IAA is better than 2,4–D

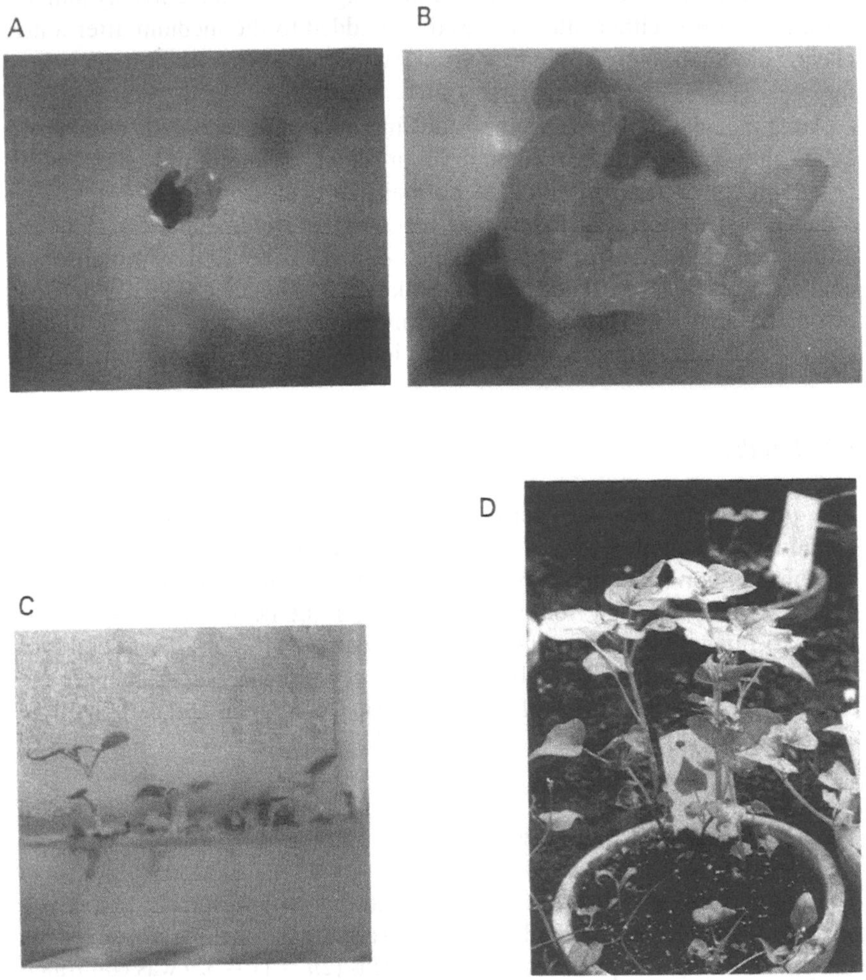


*Figure 1.* Developmental stages of ovule culture. (A) Initiation of growth. (B) Enlargement of foliar structures. (C) Regenerated shoots in culture. (D) Acclimatized plants.

at concentrations tested and that filter-sterilised 2iP at 49.20 μM is slightly better than 2iP at 24.60 μM; autoclaved 2iP exhibited much lower activity than filter-sterilised. The overall conclusion is that the basal medium supplemented with 49.20 μM filter-sterilised 2iP, 2.85 μM IAA and 90 g $l^{-1}$ maltose was an optimal induction medium.

At least 50% of the regenerants were lost during further stages of micropropagation before they reached the stable shoot culture stage. This can be attributed to genetic aberrations expressed in regenerated genotypes.

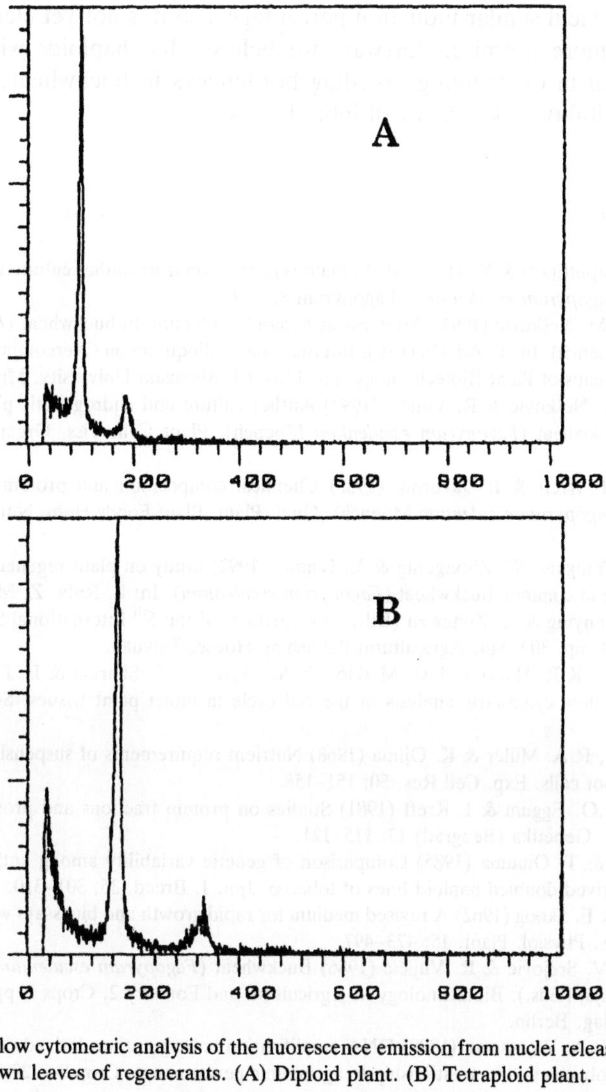


*Figure 2.* Flow cytometric analysis of the fluorescence emission from nuclei released by chopping *in vitro* grown leaves of regenerants. (A) Diploid plant. (B) Tetraploid plant.

## 4. Conclusions and prospects

Anther and ovule culture are commonly used techniques for *in vitro* production of haploids. Optimisation of procedures for haploid induction is a long term process that requires extensive testing of several parameters including status of donor plants, developmental stage of microspores/ovules, pretreatments, composition of media, culture regimes and several others. We regard the present state of haploid induction methods in buckwheat as a starting point for further experiments. At present, the protocols for both anther and ovule culture yield similar induction percentages, so it is not yet clear which method has more promise. Anyway, we believe that haploids will be an important tool in overcoming breeding bottlenecks in buckwheat, such as self-incompatibility or low vigor of inbred lines.

## 5. References


Adachi, T., S. Suputtitada & Y. Miike (1988) Plant regeneration from anther culture in common buckwheat (*Fagopyrum esculentum*). Fagopyrum 8: 5–9.

Bohanec, B. & M. Nešković (1991) Attempts at haploidy induction in buckwheat (*Fagopyrum esculentum* Moench). In: T. Adachi (Ed.), International Colloquium on Overcoming Breeding Barriers by Means of Plant Biotechnology, pp. 156–164. Miyazaki University, Miyazaki.

Bohanec, B., M. Nešković & R. Vujičić (1993) Anther culture and androgenetic plant regeneration in buckwheat (*Fagopyrum esculentum* Moench). Plant Cell. Tiss. Organ Cult. 35: 259–266.

Eggum, B.O., I. Kreft & B. Javornik (1980) Chemical composition and protein quality of buckwheat (*Fagopyrum esculentum* Moench). Qual. Plant. Plant Foods Hum. Nutr. 30: 175–183.

Fanchu, K., S. Yingkai, W. Zhongqing & Y. Linmei (1992) Study on plant regeneration from anther culture in common buckwheat (*Fagopyrum esculentum*). In: L. Rufa, Z. Ming-De, T. Yongru, L. Jianying & Z. Zongwen (Eds.), Proceedings of the 5th International Symposium on Buckwheat, pp. 309–314. Agricultural Publishing House, Taiyuan.

Galbraith, D.W., K.R. Harkins, J.M. Maddox, N.M. Ayres, D.P. Sharma & E. Firoozabady (1983) Rapid flow cytometric analysis of the cell cycle in intact plant tissues. Science 220: 1049–1051.

Gamborg, O.L., R.A. Miller & K. Ojima (1968) Nutrient requirements of suspension cultures of soybean root cells. Exp. Cell Res. 50: 151–158.

Javornik, B., B.O. Eggum & I. Kreft (1981) Studies on protein fractions and protein quality of buckwheat. Genetika (Beograd) 13: 115–121.

Kumashiro, T. & T. Oinuma (1985) Comparison of genetic variability among anther-derived and ovule-derived doubled haploid lines of tobacco. Jpn. J. Breed. 35: 301–310.

Murashige, T. & F. Skoog (1962) A revised medium for rapid growth and bioassays with tobacco tissue cultures. Physiol. Plant. 15: 473–497.

Nešković, M., V. Srejović & R. Vujičić (1986) Buckwheat (*Fagopyrum esculentum* Moench). In: Y.P.S. Bajaj (Eds.), Biotechnology in Agriculture and Forestry 2, Crops I, pp. 579–593. Springer-Verlag, Berlin.

Reed, S.M. & E.A. Wernsman (1989) DNA amplification among anther-derived doubled haploid lines of tobacco and its relationship to agronomic performance. Crop Sci. 29: 1072–1076.

Sinković, T. & B. Bohanec (1988) Chromosome counts and karyotype analysis in buckwheat (*Fagopyrum esculentum* Moench). Fagopyrum 8: 20–22.

Wernsman, E.A. (1992) Varied roles for the haploid sporophyte in plant improvement. In: H.T. Stalker & J.P. Murphy (Eds.), Plant Breeding in the 1990s, pp. 461–484. C.A.B. International, Wallingford, UK.

# 9. Haploidy in pearl millet [*Pennisetum glaucum* (L.) R. Br.]

BYUNG-HAN CHOI, KEUN-YONG PARK & RAE-KYEONG PARK

**Contents**



## 1. Introduction

Most of the current breeding in pearl millet [*Pennisetum glaucum* (L.) R. Br.] is aimed at maximum exploitation of hybrid vigour for both grain and forage yields. Pearl millet, a cross-pollinated crop, generally requires longer breeding periods in developing pure lines and hybrids compared with self-pollinated crops. Haploids in pearl millet would help produce homozygous lines in a single step. By developing elite breeding lines in a shorter period, breeding programmes can be accelerated. Haploids are also useful for studying mutagenesis, gene dosage effects, and interaction and linkage of genes. Hybrid seed production and maintenance of parental inbred lines require more labour and larger acreage than for self-pollinated crops due to the need for emasculation to prevent self-pollination and isolation of hybrid seed production fields. Because of its potential for improvement and development of new breeding materials, anther or microspore cultures are currently under investigation.

The objectives of this paper are to indicate the importance of pearl millet as a crop and to address the progress and problems in using anther culture, microspore and tissue culture in haploid production in this species.

## 2. Pearl millet

Pearl millet is a diploid ($2n = 2x = 14$) $C_4$ plant and summer crop that originated from West Africa. It is the sixth most important cereal in the world and widely cultivated in the semiarid tropics as a major staple food crop. It is grown on an estimated 28 million hectares worldwide. It is a hardy cereal suited to areas with low and erratic rainfall with sandy infertile soils, such that sorghum and maize production are prohibited. The grain is used to make unleavened bread (chapatis) in south Asia or prepared as gruel,

*S.M. Jain, S.K. Sopory & R.E. Veilleux (eds.),* In Vitro *Haploid Production in Higher Plants, Vol. 4,* 171–179.

*Table 1.* Forage yield performance of pearl millet hybrid Chungaecho (Suwon, 1986–87)

| Crop (Hybrid) | Forage yield (t/ha) | Dry matter percentage (%) | Dry matter yield (t/ha) | Leaf area index | Highest forage yield** (t/ha) |
|---|---|---|---|---|---|
| Pearl millet (Chungaecho) | 150 | 17.7 | 28 | 28.8 | 192 |
| Maize (Suwon 19)* | 66 | 31.8 | 21 | 5.8 | 83 |
| Sorghum/sudan grass (GW9110) | 111 | 28.2 | 27 | 19.5 | 143 |

* Maize harvested 40 days after silking (1986).
** '87 highest forage yield harvested from the early planted, heavily fertilized and polyethylene film mulched plot.

dumplings, couscous and beer in Africa. It is also used as an animal feed and forage in both temperate and tropical regions due to its ability to thrive in poor and dry soil (Andrews, 1986). Its forage yield compared favourably with that of maize and sorghum (Table 1).

Pearl millet is either equal or superior to wheat, maize, sorghum, and rice grain protein and oil content. If pulse is the poor man's meat, millet is the poor man's bread. Millet consumers are strong and healthy. Rats fed on different unsupplemented diets of either pearl millet, ragi, sorghum or maize, demonstrated the highest growth rate on pearl millet (Burton *et al.*, 1972; Rooney and McDonough, 1987).

Pearl millet is a robust, annual bunchgrass with extreme variation. A plant may consist of a single culm less than 50 cm in height to many culms up to 5 m tall (Fig. 1). Culms may be simple or branched, slender or stout, smooth or hairy. Leaves and sheaths may also be smooth or hairy. Plants parts may be green, purple, red, or golden yellow. Today the greatest morphological diversity in pearl millet occurs in west Africa, south of the Sahara Desert and north of the forest zone. The wild progenitor also occurs in the drier northern portion of this zone. Present day distribution indicates that the Sahel zone extending from western Sudan to Senegal is the original home of pearl millet (Burton, 1980). The International Crops Research Institute for the Semi-Arid Tropics (ICRISAT) in India is presently centralizing and enlarging the world collection and developing superior cultivars and hybrids for the semi-arid tropic regions.

The inflorescence of pearl millet is a false spike ranging from 5 to more than 150 cm in length and from 1 to 5 cm in diameter. An important feature of the floral biology of pearl millet is its conspicuous protogynous nature (Fig. 2). The protogynous flowering habit facilitates cross pollination. Stigmas generally emerge on flowers near the tip of the partially exserted spikes and dehiscence proceeds downward. The first stamens appear 2–3 days after the stigmatic branches for pearl millet growing under arid conditions. Complete emergence of stigmas on a spike generally occurs within 2 days. However,

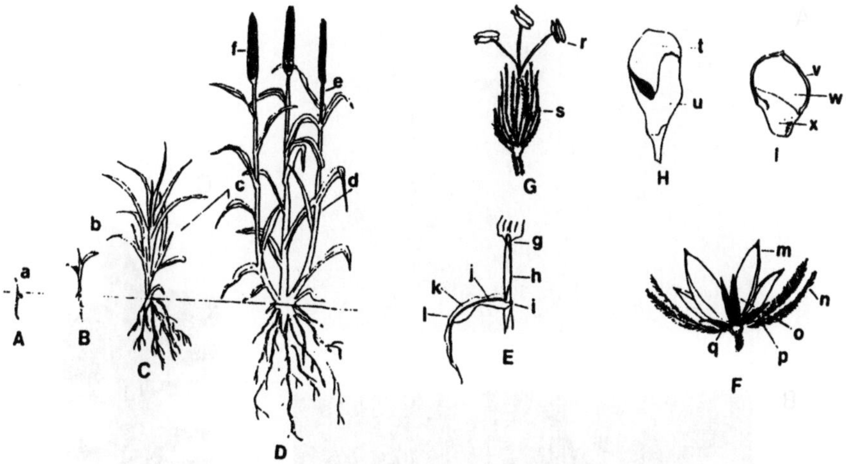


*Figure 1.* Diagrammatic stages of the life cycle of a pearl millet plant, illustrating plant parts. A = seedling; B & C = vegetative stages; D = adult plant; E = part of stem; F = spikelet; G = single flower; H = spikelet and grain; I = grain; a = coleoptile; b = foliage at vegetative stage; c = foliage at vegetative and adult plant stage; d = stem; e = peduncle; f = earhead; g = ligule; h = leaf sheath; i = node; j = midrib; k = leaf blade; l = leaf margin; m = upper perfect flower; n = bristles subtending group of spikelets; o = upper glume; p = lower staminate flower; q = lower glume; r = penicilled anther; s = group of two spikelets subtended by several bristles; t = grain; u = glume; v = pericarp; w = endosperm; x = embryo (Kumar & Andrews, 1993 (Crop Sci.)).

this depends both on the length of the spike and environmental conditions. Self pollination as high as 31% has been reported (Burton, 1974).

The first anthers generally emerge on apical florets approximately one day after most styles on a head have excised. Most heads shed pollen for 4 to 6 days. When temperatures exceed 25°C, anthesis occurs anytime during the day with the greatest flush of anthers appearing soon after sunrise. Fertilization occurs within a few hours after pollination (Burton, 1980).

Kumar & Andrews (1993) reviewed the literature on genetics of qualitative traits in pearl millet. The number of morphological mutants reported was 145 including chlorophyll deficiencies (26%), plant pigmentation (18%), and earhead characters (14%). Only a few mutant traits (e.g., cytoplasmic nuclear male sterility, the dwarfing gene *d2*, and the trichomeless gene (*tr*) may be of significance in the improvement of grain and forage production. Other mutant traits including brown midrib have potential for increasing forage quality.

### 3. Heterosis breeding in pearl millet

Pearl millet is ideally suited for exploitation of heterosis because of its high level of heterozygosity and susceptibility to inbreeding depression. Efforts

A

B

C

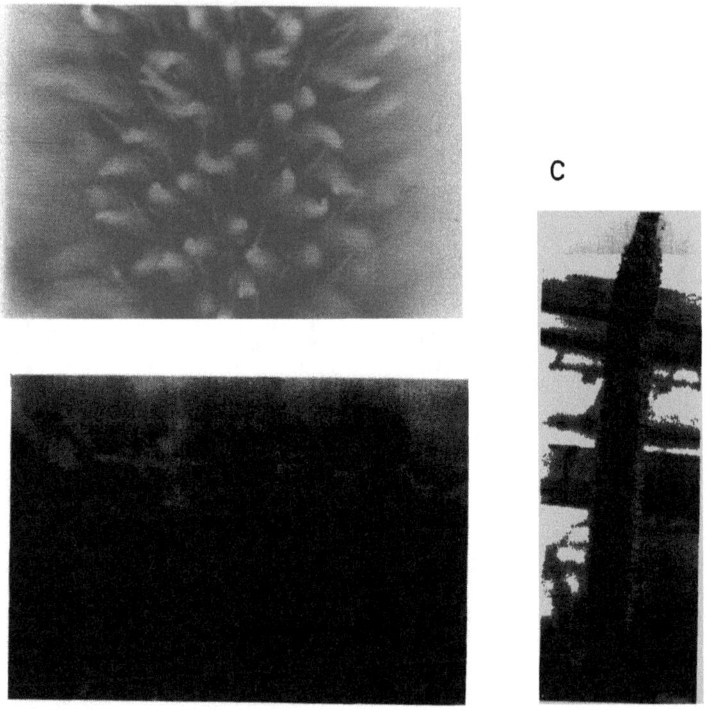

*Figure 2.* Conspicuous protogynous nature of flowering habit in pearl millet facilitates cross pollination. (A) Stigmas of staminate flower; (B) stigmas and penicilled anthers; (C) complete emergence of stigmas, and partially emerged anthers shedding pollen on one half of a spike grown in the greenhouse of Crop Experiment Station, Suwon, Korea in 1994 winter and spring.

have been concentrated on heterosis breeding for both grain as well as forage production. Inbreeding depression is probably mainly the result of uncovering deleterious recessive alleles which normally remain masked in the heterozygous state in allogamous crops. Different genotypes have been found to show a differential responses with reagrd to inbreeding depression. Significant heterosis has been observed in pearl millet yield (Choi *et al.*, 1993) (Table 2). The greater photosynthetic surface coupled with higher photosynthetic efficiency apparently contributed to a greater biomass production in the hybrid. The length of grain filling period and the rate of grain filling were correlated with final dry grain weight, and grain filling was faster in hybrids than in the parental lines.

Inbreeding of parental lines T186, T23DA, and T23DB led to a sudden decrease in vigour, and exposed six recessive types of chlorophyll deficiencies and four types of recessive sterility. A marked inbreeding depression also was observed for several agronomic traits including plant height, number of tillers, head weight and yield, but intermediate for several other characters.

*Table 2.* Major agronomic traits of parental lines of Chungaecho hybrid of pearl millet (Suwon, 1986)

| Parental lines and their hybrid | Heading date | Plant height (cm) | Tillers per plant | Stem diameter (mm) | Spike | |
|---|---|---|---|---|---|---|
| | | | | | Length (cm) | Diameter (mm) |
| T23 DA (♀) Seed parent | Aug. 6 | 134 | 7.4 | 13.1 | 17.0 | 16.8 |
| T23DB (♂) Maintainer of T23DA | Aug. 6 | 132 | 6.0 | 13.0 | 12.5 | 14.7 |
| T186 (♂) Pollen parent | Aug. 18 | 356 | 4.9 | 13.4 | 26.1 | 20.1 |
| Chungaecho hybrid | Aug. 10 | 430 | 11.0 | 15.6 | 30.7 | 18.8 |

Enforced inbreeding also lead to meiotic irregularities including a decline in chiasma frequency (Jauhar, 1981). Heterozygotes produced by crossing inbred lines show significant heterosis for chromosome behavior including increased chiasma frequency. Heterosis for both forage and grain yield closely parallels the heterozygosity of the breeding material.

The diverse germplasm in pearl millet is useful to create usable genetic variability and to broaden the genetic bases of cultivars. ICRISAT's centralized world collection of diverse germplasm should serve pearl milllet breeders who seek to improve agronomic traits or better adaptation of the crop.

## 4. Haploids in pearl millet

A systematic search for haploids was initiated in pearl millet after discovery of a haplo-haplo-twin seedling. Powell *et al.* (1975) studied thirteen haploid plants ($2n = 1x = 7$) of pearl millet. Two plants resulted from a twin embryo caryopsis, whereas the others were discovered in field plantings. The rate of spontaneous haploid occurrence was 1 per 10,000 plants in the inbred Tift23A. Haploids were conspicuous because of their small inflorescence diameters, narrow leaves, and poor exsertion and dehiscence of anthers. Meiotic chromosomes of haploids were mostly univalents, but occasional pairing of two or more chromosomes was observed at metaphase I. Seed set for haploid plants ranged from none to 502 compared with the usual seed set for diploid plants ranging from 1,000 to 3,000. No haploids were recovered among progeny of seeds harvested from haploid plants. Spontaneous chromosome doubling occurred in several of the haploid plants. Doubled sectors in the inflorescences were characterized by good exsertion and good dehiscence of anthers. The tendency in pearl millet for haploid plants to double their chromosome number spontaneously can be exploited to develop autoploid lines without colchicine treatment (Powell *et al.*, 1975).

The potential of anther culture for genetic fixation and selection of promising progeny lines derived from hybrid populations has long been realized.

*Table 3.* Response in anther culture of pearl millet subjected to various low temperature pretreatments (Suwon, 1989–1992)

| Temperature | Duration | Anthers plated | Embryoids induced (%) | Plants regenerated | |
|---|---|---|---|---|---|
| | | | | Green | Albino |
| Control | – | 5,050 | 7 (0.14) | 0 | 0 |
| 9°C | 9 days | 6,500 | 15 (0.23) | 4 | 1 |
| | 14 days | 5,800 | 21 (0.36) | 6 | 1 |
| 14°C | 9 days | 6,500 | 18 (0.28) | 5 | 1 |
| | 14 days | 5,000 | 8 (0.14) | 2 | 0 |

Haploids obtained from anther culture of hybrid plants can be doubled through various means to generate inbred lines that recombine the genes of parental lines used to create the hybrid. Anthers of hybrid plants can be cultured to produce doubled haploids. The pure lines developed from culture must carry desirable genes from parental lines for producing superior hybrids or synthetic cultivars. The breeding cycle may be shortened by 5 to 6 generations when the haploid breeding method is used along with conventional cross breeding methods. The transfer of valuable foreign genes of landraces into cultivars to enrich the genetic background of cultivars necessary for crop improvement may also be facilitated through anther culture.

Ha & Fernes (1982) were the first to report successful regeneration of haploids from a pearl millet line (Tift 23 $D_2$ $B_1$) and an $F_1$ hybrid Massue × Liqui through anther culture. Since their original report, significant progress has been made in embryo and callus production from cultured anthers (Choi *et al.*, 1989).

The growth of the donor plants and developmental stages of the microspores have been important factors in the initiation of regenerable anther cultures. To sample and culture optimal aged spikes, pearl millet stems should be cut well below the uppermost leaves including flag leaf before the spike has emerged, and the upper portion of the plant brought to the laboratory. Higher success rate of anther cultures has been observed when the detached spikes were exposed to a low temperature pretreatment for 9–14 days and the anthers were osmotically stressed in sucrose solution. The optimal pretreatment temperature was 9°C for 14 days (Table 3). The stem should then be carefully opened by splitting it along its length, taking care not to damage the spike which should be removed intact and kept in a moist environment.

It has been convenient to draw a diagram of the spike, sample florets from various portions of the spike, and note on the diagram the development stage of the micropspores from the different locations and then decide which portions of the spike to use for anther culture. It is possible to stain microspore nuclei with acetocarmine stain with 50 mg Tryan blue in 100 ml distilled water. The cells can be examined immediately and the stage quickly identi-

*Table 4.* Osmotic pressure stress for anther culture of pearl millet hybrid (Suwon, 1989–1992)

| Pretreatment | Anthers plated | Embryos induced (%) | Plants regenerated | |
|---|---|---|---|---|
| | | | Green | Albino |
| Osmotic pressure | 2,600 | 24 (0.92) | 7 | 3 |
| Control | 2,550 | 10 (0.39) | 2 | 1 |

*Table 5.* Medium for pearl millet anther culture (Suwon, 1989–1992)

| Medium | Basic medium | Supplements |
|---|---|---|
| PACM 1 | N6 | 500 mg/l casein hydrolysate + 2 ppm 2,4–D + 1 ppm kinetin + 2 ppm BA + 15% sucrose + 0.5% active carbon + 0.8% agar |
| PACM 2 | Yu-Pei | 500 mg/l lactoalbumin hydrolysate + 0.1% TIBA + 12% sucrose + 0.5% active carbon + 0.8% agar |

*Table 6.* Anther culture response of pearl millet (Suwon, 1989–1992)

| Medium | Anthers plated | Embryos induced (%) | Plants regenerated | |
|---|---|---|---|---|
| | | | Green | Albino |
| PACM 1 | 24,450 | 38 (0.16) | 4 | 1 |
| PACM 2 | 24,300 | 55 (0.23) | 7 | 1 |

fied. The optimal stage of pearl millet anther culture development has been found to be when the microspores are uninucleate and the nucleus is located in the center of the microspore and when the spike is still enclosed in the flag leaf sheath (Choi *et al.*, 1989). Osmotic pressure and specific gravity treatment by centrifugation at 1,000 rpm for 10 min in 25% sucrose solution were effective for successful anther culture (Table 4). When anthers were put into higher sucrose solution and then centrifuged, both embryo formation and plant regeneration rate increased.

For anther culture of pearl millet, uninucleate microspores enclosed in the anthers were placed on N6 (Chu *et al.*, 1975) and Yu-Pei (Ku *et al.*, 1978) basic media with supplements as shown in Table 5. The cultures were then incubated at 28 ± 1°C in the dark for 4 weeks and finally transferred to high fluorescent light at 16 h photoperiod for incubation. A low frequency of embryos regenerated into green plants (Tables 3, 4 and 6). The embryos should be subcultured frequently to maintain vigorous growth, at 2 week intervals for fast growing cultures. Maintenance of the subcultures and incubation conditions must be followed by careful management through the improved culture techniques. There were significant differences in response to anther culture of pearl millet due to genotype × medium and environmental interactions (Choi *et al.*, 1989). A large number of plants could be induced from elongating internodes of pearl millet × napier grass hybrid. Potentially, clones of these vigrous hybrids may be used for forage production.

## 5. Conclusion and prospects

Pearl millet is a hardy cereal suited to dry regions with sandy infertile soils where rainfall is low and erratic, but also has the capability to grow well in fertile soils. Pearl millet grain is equal or superior to wheat, maize, sorghum, and rice in protein and oil content. Haploid plants of pearl millet are useful for the development of new cultivars. The breeding cycle could be shortened by five to six generations when haploid breeding method is used along with conventional cross breeding methods. Anther culture is useful for creating the genetic diversity of germplasm in pearl millet breeding programmes. However, the frequency of haploid induction in pearl millet has been very low. Modifications to the medium or improved techniques must be developed to improve the performance of pearl millet anther culture in the future.

## 6. Acknowledgements

The authors would like to thank scientists and technologists of Crop Experiment Station for cooperation and assistance in conducting anther and tissue culture of pearl millet. My sincere gratitude is extended to Docent Dr. S. Mohan Jain of Plant Production Department, University of Helsinki for providing me information on editing for a book entitled "*In Vitro* Haploid Production in Higher Plants".

## 7. References

Andrews, D.J. (1986) Breeding pearl millet grain hybrids. University of Nebraska-Lincoln: 1–23.

Burton, G.W., W.A. Wallace & K.O. Rachie (1972) Chemical composition and nutritive value of pearl millet grain. Crop Sci. 12: 187–188.

Burton, G.W. (1980) Pearl millet: In W.R. Fehr & H.H. Hadley (Eds.), Hybridization of Crop Plants, pp. 457–469. American Society of Agronomy-Crop Science·Society of America, Madison, WI.

Burton, G.W. (1974) Factors affecting pollen movement and natural crossing in pearl millet. Crop Sci. 14: 802–805.

Choi, B.H., K.Y. Park & R.K. Park (1989) Response to anther and tissue cultures of corn, pearl millet and buckwheat genotypes. Korean J. Crop Sci. 34: 142–146.

Choi, B.H., K.Y. Park & R.K. Park (1993) Pearl millet hybrid of high quality and yield: A new green fodder crop in Korea. In: K.J. Kim et al. (Eds.), Crop Production and Improvement Technology in Asia, pp. 485–497. KSCS, Korean Society of Crop Science.

Chu, C.C., C.C. Wang, C.C. Sun, C. Hsu, K.C. Yin, C.Y. Chu & F.Y. Bi (1975) Establishment of an efficient medium for anther culture of rice through comparative experiments on the nitrogen sources. Sci. Sin. 18: 659–668.

Ha, B.D. & J. Fernes (1982) Androgenesis in pearl millet. I. Analysis of plants obtained from microspore culture. Z. Pflanzenphysiol. 108: 317–327.

Jauhar, P.P. (1981) Cytogenetics and Breeding of Pearl Millet and Related Species, pp. 71–75. Alan R. Liss Inc., New York, NY.

Ku, M., W. Cheng, L. Cuo, Y. Kuan, H. An & C. Huang (1978) Induction factors and morpho-cytological characterestics of pollen-derived plants in maize (*Zea mays*). In: Proceedings of Symposium on Plant Tissue Culture, pp. 35–42. Science Press, Peking.

Kumar, K.A. & D.J. Andrews (1993) Genetics of qualitative traits in pearl millet: a review. Crop Sci. 33: 1–20.

Powell, J.B., W.W. Hanna & G.W. Burton (1975) Origin, cytology, and reproductive characteristics of haploids in pearl millet. Crop Sci. 15: 389–392.

Rooney, L.W. & C.M. McDonough (1987) Food quality and consumer acceptance of pearl millet. In: Proceedings of the International Pearl Millet Workshop held on April 7–11, 1986, ICRISAT Center, International Crops Research Institute for the Semi-Arid Tropics, pp. 43–61.

# 10. Haploidy in rye

SABINE DEIMLING and TANJA FLEHINGHAUS-ROUX

**Contents**



## 1. Introduction

Rye (*Secale cereale* L.) has its origin in southwest Asia. From there it was probably distributed to Russia, and later to western Europe. Only 2–3% of total cereal production falls to rye. Approximately 90% of rye production is concentrated in Europe. The distribution of rye is restricted to the area between the 50th and 60th degree of northern latitude. Due to its relative tolerance for a range of soil and climatic conditions, *S. cereale* is grown mainly in northern regions, where other cereals often fail. In southern regions, rye competes with wheat as bread grain for human nutrition. Today, the main rye growing area is eastern Europe with approximately 14 million hectares per year. In 1993 the EU produced 4 million tons of rye on slightly over 1 million hectares. Within the EU countries, Germany is the major rye producer with 3 million tons on 650 000 hectares.

Rye is an open-pollinating species with a gametophytic self-incompatibility system. In contrast to wheat, barley or oats, where homozygous lines are released as cultivars, rye cultivars are highly heterozygous. Until 20 years ago, breeding progress in *S. cereale* was achieved through the continuous development of either open-pollinated or synthetic cultivars. A notable step forward in breeding efficiency took place after some biological peculiarities were discovered: the self-incompatibility mechanism can be neutralized by

*S.M. Jain, S.K. Sopory & R.E. Veilleux (eds.),* In Vitro *Haploid Production in Higher Plants, Vol. 4,* 181–204.

genes for self-fertility. The incorporation of such genes enables the breeder to produce inbred lines through continuous selfing (Geiger, 1982). The "Pampa" cytoplasm was discovered representing a source of cytoplasmatic male sterility (cms) which facilitates crossing among inbred lines (Geiger & Schnell, 1970). Fertility can be restored by the introduction of restoration genes into the pollinating parent (Geiger, 1972). These prerequisites enable exploitation of heterosis in rye through breeding of hybrid cultivars. The first hybrid cultivars were released in 1984 and since then have become increasingly important in commercial rye breeding (Geiger, 1990). The major advantages of hybrid over open-pollinated cultivars are increased yield of about 10–15% and improved lodging resistance due to shorter straw. In 1993 the acreage of hybrid cultivars was 15% in the EU, but had already reached 43% in Germany.

The production of haploids and subsequent development of doubled haploid lines (DHL) have proven valuable in plant breeding and biotechnology. The incorporation of DHL in breeding is advantageous for several reasons: the production of DHL hastens the development of homozygous lines by 4–5 generations compared to inbred lines developed conventionally by selfing. However, in hybrid breeding, testing for combining ability is the most time consuming process. This cannot be reduced by the DHL approach. Since conventional breeders usually start testcrossing in early selfing generations, only one or two years can be saved in the hybrid breeding process, especially in winter cereals like rye. Nevertheless, for special breeding purposes, where unselected lines are desired, full advantage can be taken of the abovementioned time saving procedure. Using DHL, the increased genotypic variance of the line *per se* and its testcross performance lead to a better differentiation of quantitative traits. Consequently, this results in higher and more predictable gain from selection (Geiger, 1985). The uniformity of DHL facilitates plant variety protection. This will become more important when even greater homogeneity of cultivars will be demanded by governmental regulations.

While DHL are of interest in commercial plant breeding, they are also valuable for scientific purposes related to plant breeding. In mapping studies, DHL are especially useful because they are true-breeding. This allows large scale evaluations, which result both in increased accuracy of character assessments and also in higher efficiency of mapping quantitative trait loci (QTL). Depending on heritability, under certain circumstances, DHL can be used more efficiently as a mapping population than $F_2$ or $F_2$-derived populations (Melchinger, 1990). In transformation experiments, microspores are attractive as recipients of foreign DNA and might become more important for genetic transformation of cereals. Recently, successfully transformed (unicellular) barley microspores have been regenerated into fertile, completely homozygous plants, yielding transgenic DHL (Jähne *et al.*, 1994).

The practical use of DHL in rye breeding needs, as the basic requirement, the production of large numbers of doubled haploids without serious genotypic limitations and without extreme expenditures. As in other species, there are several possibilities in rye to produce haploid and doubled haploid plants. Using interspecific pollination, haploid embryos can be produced via amphimixis. By pollinating rye with maize pollen, Laurie *et al.* (1990) observed a fertilization rate of 18.7%; however, embryos and plants were not obtained. Using the same technique, Deimling *et al.* (1994) obtained six embryos from which two DHL resulted after pollination of 48 000 emasculated flowers. One was induced after pollination with pearl millet, the other with maize. The effort required for emasculation and pollination of such a large number of rye flowers to regenerate so few haploids is far too great for interspecific pollination to be a suitable haploidization technique in rye. Some *in vitro* techniques (ovary, anther, and microspore culture) appear more promising. The utilization of ovary culture for rye haploid production has either never been investigated or its failure has not been published. In contrast, anther and microspore culture have been the subject of several investigations over the past 20 years. The present chapter focuses on this work.

## 2. Anther culture

### 2.1. *Literature overview*

The relatively few publications (Table 1) on rye anther culture compared to other cereals may reflect both the lesser importance of rye worldwide and its general recalcitrance towards tissue culture techniques. In general, scientific work on rye is being conducted only in those countries where this cereal is grown as bread grain for human nutrition (Seibel, 1988). The main publications on androgenic haploidy in rye, including the genetic material used, the factors investigated, and the abbreviated results are given in Table 1.

### 2.2. *Factors influencing androgenic capacity*

Six factors can be recognized that affect androgenesis and subsequent DHL production. These are: 1) physiological condition of the donor plants; 2) stage of microspore development at the time of culture; 3) spike and/or anther pretreatment; 4) composition of culture medium; 5) conditions during culture; and 6) the genotype (perhaps most important). Some of these factors are more important for the induction of androgenic structures whereas others influence the development of androgenic structures into plants. All six factors have been investigated in relation to rye anther culture, as noted in Table 1. Throughout the following discussion, the induction rate is given as the number of responding anthers (anthers showing any kind of structures) compared to the number of anthers plated (excluding contaminated cultures).

*Table 1.* Literature overview on *in vitro* haploidy in rye, including the genetic background of donor plants (*Secale cereale* (*S.c.*), *S. vavilovii* (*S.v.*), *S. montanum* (*S.m.*)), factors investigated (1 = state of donor plant, 2 = anther/microspore developmental stage, 3 = spike/anther pretreatment, 4 = culture medium, 5 = culture conditions, 6 = genotype), and success in anther culture response (induction/regeneration + or −)

| Author | Year | Material | Investigated factors | Anther culture response | |
|---|---|---|---|---|---|
| | | | | Induction | Regeneration |
| Zenkteler & Misiura | 1974 | *S.m.* | 4 | + | − |
| Wenzel & Thomas | 1974 | *S.c.* | 2, 4 | + | + |
| Thomas *et al.* | 1975 | *S.c.* × *S.v.* | 1, 2, 3, 4 | + | + |
| Wenzel *et al.* | 1976 | *S.c.*, *S.c.* × *S.v.* | 6 | + | + |
| Wenzel *et al.* | 1977 | *S.c.*, *S.c.* × *S.v.* | 3, 4, 6 | + | + |
| Orlikowska | 1977 | *S.c.* | 4 | + | − |
| Hoffmann & Wenzel | 1981 | *S.c.*, *S.c.* × *S.v.* | agronomic value of DHL | + | + |
| Sharma *et al.* | 1982 | *S.c.* | 3, 4, 6 | + | − |
| Friedt *et al.* | 1983 | *S.c.* × *S.v.* | agronomic value of DHL | + | + |
| Lind *et al.* | 1985 | *S.c.*, *S.c.* × *S.v.* | agronomic value of DHL | + | + |
| Wenzel *et al.* | 1985 | *S.c.*, *S.c.* × *S.v.* | agronomic value of DHL | + | + |
| Milewska-Pawliczuk | 1987 | *S.c.* | 4, 6 | + | − |
| Lorenz | 1989 | *S.c.*, *S.c.* × *S.v.* | 2, 4, 5 | + | − |
| Flehinghaus *et al.* | 1991 | *S.c.*, *S.c.* × *S.v.* | 3, 4, 6 | + | + |
| Hörlein | 1991 | *S.c.*, *S.c.* × *S.v.* | 1, 4, 5, 6 | + | + |
| Deimling *et al.* | 1992 | *S.c.* × *S.v.* | 4 | + | + |
| Daniel | 1993 | *S.c.*, *S.c.* × *S.v.* | 4, 6 | + | + |
| Flehinghaus-Roux | 1994 | *S.c.*, *S.c.* × *S.v.* | 1, 3, 4, 6 | + | + |
| Flehinghaus-Roux *et al.* | 1995 | *S.c.*, *S.c.* × *S.v.* | 6 | + | + |

Regeneration rate is given as the number of plants regenerated per number of anthers plated (again excluding contaminated cultures).

### 2.2.1. *Growth of donor plant material*

Survival and response of the microspores have been maximized by growing the plants used as a source of androgenic cultures under optimal conditions. In our experience, field grown rye is unsuitable due to a high contamination rate (Hörlein, 1991). The vernalization period of the donor plants is important for regeneration of androgenic plants. Best anther culture response has been observed after plants were vernalized at 2°C for 8 to 14 weeks with an optimum at 10 weeks (Flehinghaus *et al.*, 1991). Vernalization for less than 8 weeks resulted in weak donor plants with few spikes and poor anther culture response. Different genotypes require distinct lengths of vernalization but, in general, longer durations are advised. Daniel (1993) vernalized his material for 12 weeks at 4° C. Best results have been obtained in our lab using the following protocol:

- Sow the seeds in Jiffy-pots (3 cm, Jiffy, Norway).
- Germinate at 20/15°C (day/night) with 16 h light (15 000 lux) for one week.
- Vernalize at 2–leaf stage at 2°C under 8 h daylength and dim light (approx. 1 000 lux) for 10 weeks.
- Adaptation-period at 14°C under 10 h higher light (approx. 2 000 lux) for 2 weeks.
- Transfer to soil in the greenhouse (20°C day/15°C) night with a minimum daylength of 16 h provided by high pressure mercury lamps giving at least 15 000 lux.

The adaptation period under intermediate temperatures was beneficial for tillering of donor plants. Daniel (1993) successfully applied a tillering period at 10/8°C (day/night) before plants were transferred to 18/14°C (16 h photoperiod) greenhouse conditions. Thomas *et al.* (1975) recommended donor plant growth at 20/18°C (16 h light) with at least 6 000 lux light intensity measured at the top of the plants. Satisfactory results could be obtained by growing the donor plants in either a greenhouse or a growth chamber (Flehinghaus-Roux, 1994). When greenhouse temperature could not be controlled during summer (May until September) and day temperature reached 25–30°C, induction and regeneration rates dropped dramatically (Fig. 1). Previous experiments with an artificial environment led to the growth chamber conditions which were routinely applied parallel to the greenhouse experiments: temperature ranging between 16 and 20°C, daylength of 16 h provided by high pressure mercury lamps giving 80 000 lux. Although these growth chamber conditions were kept constant over one year, large seasonal differences in anther culture response were obtained: while the induction rates varied only slightly from month to month, the regeneration rates dropped dramatically during the summer, as did that of anthers taken from greenhouse material (Fig. 1). Except during early spring, anther culture results were generally slightly better using material from the growth chamber. Using greenhouse material, good results were obtained from fall through spring, but it seems advisable to stop anther culture experiments during the hottest months of the year.

Donor plant growth under low temperature conditions as described for other cereals (Foroughi-Wehr & Mix, 1979) did not lead to satisfactory results in rye. Cool temperature (12°C) combined with low light intensity (less than 15 000 lux) resulted in poor plant development and consequently low anther culture response (Hörlein, 1991). Treatment of plants with topical or systemic pesticides should be avoided during donor plant growth until harvest of the spikes. Pesticide treatment (or the disease itself) led to a drastic drop in anther culture response. Donor plants should be quarantined to reduce fungal problems – in most of our experiments it was possible to grow donor plants without pesticides.

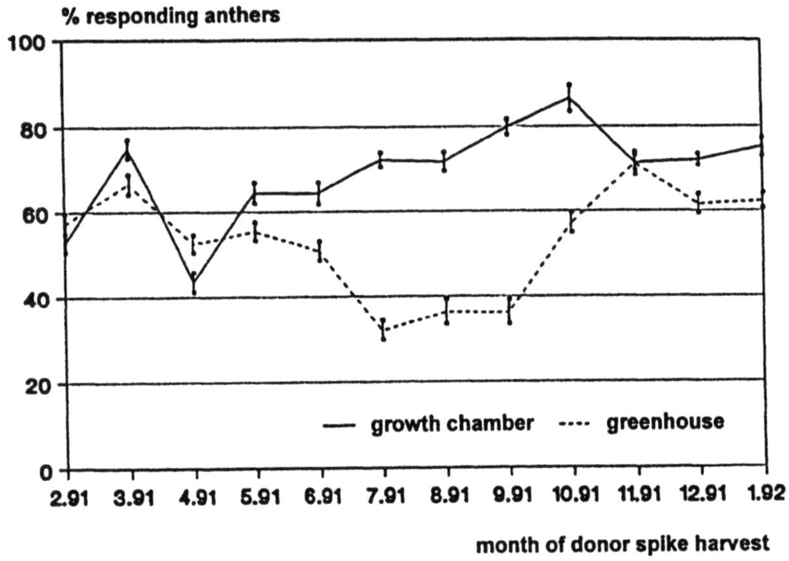


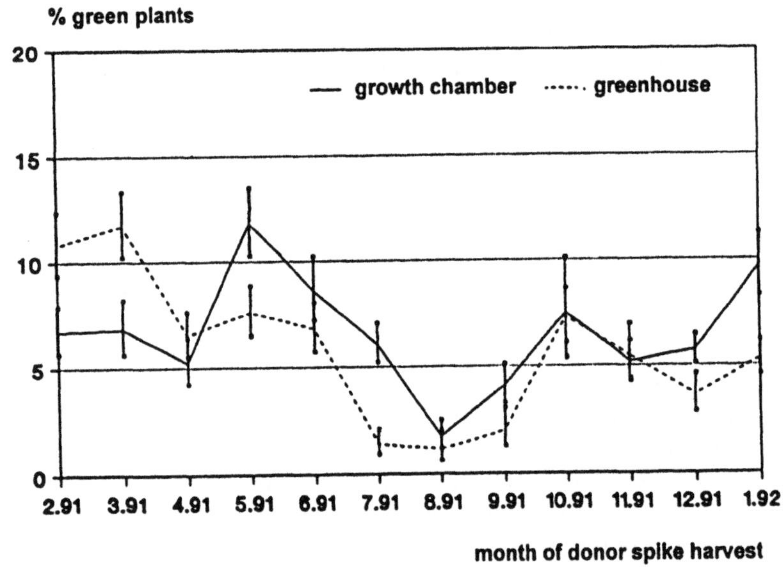


*Figure 1.* Induction (upper part) and regeneration rates (lower part) in anther cultures of single-cross rye SC35 (DH3 × DH5) after donor plants were grown in two different environments over one year (≥400 anthers plated per week and environment; vertical bars indicate the 95% confidence interval).

### 2.2.2. *Developmental stage of microspores and preconditioning*

The optimal harvest date for donor spikes depends on the genotype, the environmental conditions, and the season of the year. The right time for cutting spikes is around heading. Nevertheless, it is important to observe the developmental stage of the microspores in each spike before anther culture. Best results have been obtained when the majority of anthers contain microspores at the mid- to late uninucleate stage of development. This is when the nucleus starts migrating from the wall near the pore towards the center of the microspore. To estimate the pollen stage, one flower is excised from the center of the spike, its microspores stained with aceto-carmine, and the position of the nucleus determined. For rating, the description of the six pollen grain stages (1 = tetrad stage, 6 = first pollen nuclear division) published by Gaul *et al.* (1976) was followed. Only those spikes with microspores rated 4 or 5 according to Gaul *et al.* (1976) should be selected. Lorenz (1989) and Daniel (1993) found large genotypic differences for the most suitable microspore stage and recommended the early binucleate stage for anther culture of some genotypes.

Another factor that may enhance the induction of androgenesis is the application of a temperature shock. This stimulation can either be applied as a pretreatment before plating anthers or directly after plating. The kind of shock (heat or cold, optimal temperature) and treatment (pre- or post-plating, duration) is species dependent and often needs to be adapted even to single cultivars. For maize, a combination of a pre- and post-plating cold treatment was described by Dieu & Beckert (1986), Deimling *et al.* (1990), and Büter *et al.* (1991).

For rye, a cold treatment to the spikes directly after harvest has routinely been used. Thomas *et al.* (1975) preferred 6°C for 3–15 days, Lorenz (1989) applied 4–5°C for 8–14 days, whereas Daniel (1993) followed our protocol of 4°C for seven days (Flehinghaus *et al.*, 1991). In contrast, Orlikowska (1977) did not observe an increase in the number of embryogenic microspores after a 1–2 day spike treatment at 4°C. Investigations on the influence of a post-plating temperature treatment in addition to the cold pretreatment in rye clearly showed an increase in anther culture response (Deimling *et al.*, 1990; Flehinghaus *et al.*, 1991). The increased response after application of cold treatments was not due to higher induction rates, but rather to improved regeneration rates (Fig. 2). During induction, no significant difference was observed for the seven day treatments at 14°C or 18°C compared to the untreated control at 27°C. However, with genotype DH5, regeneration rates were clearly increased in both the 14°C and 18°C treatments. In contrast, a seven day incubation at 8°C negatively affected both induction and regeneration rates of this genotype compared to the 27°C control. Because other genotypes yielded an optimal response after a 14°C post-plating incubation or at even lower temperatures, we routinely apply a post-plating temperature treatment for one week at 14°C for all genotypes. This post-plating treatment was beneficial in addition to pre-incubation of spikes (stalks in water) at 4°C

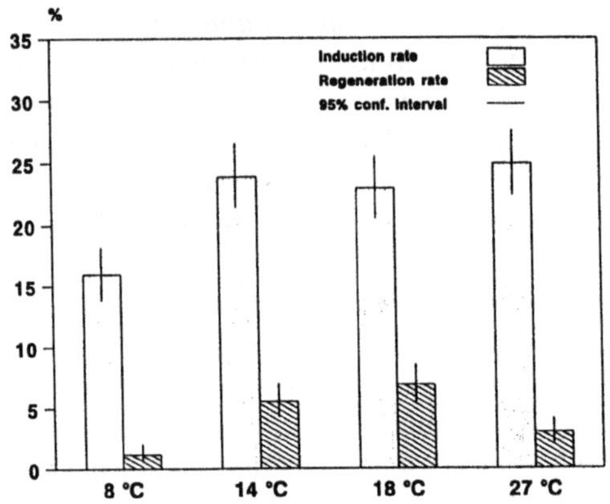


*Figure 2.* Influence of three different 7-day post-plating temperature treatments compared to direct incubation at 27°C on the induction (% responding anthers) and regeneration (% green plants) of rye genotype DH5 (vertical bars indicate the 95% confidence interval).

for 7 days in the dark. In contrast, Sharma *et al.* (1982) did not find significant differences after a post-plating treatment of 5°C for 64 h compared to incubation at room temperature. This discrepancy can be understood, if one considers that the optimal temperature treatment depends on the genotype (Marsolais *et al.*, 1984; Powell, 1988).

2.2.3. *Culture media and culture conditions*

Various basal media have been used for rye anther culture. Wenzel *et al.* (1977) suggested the use of potato extract medium (Anonymous, 1976). Because of the strong influence of the potato cultivar and quality as well as the rather complicated and laborious preparation, this medium seems inadvisable. Wenzel *et al.* (1977) also recommended the utilization of a modified N6 medium (Chu, 1978) that was initially developed for rice anther culture. We have used this medium routinely in our experiments (Flehinghaus *et al.*, 1991). Other authors, such as Daniel (1993), favoured a modified MS medium (Murashige & Skoog, 1962; Olsen, 1987).

For rye, higher androgenic capacity has been accomplished by replacing sucrose with maltose as carbohydrate source in the induction medium (Flehinghaus *et al.*, 1991). This substitution increased both the induction and regeneration rates (Table 2). With DH 5, the substitution caused a four-fold increase in the induction rate and a ten-fold increase in the number of green plants. The positive effect of maltose was evident but not as great for "Teku". The carbohydrate concentrations used in this investigation (Table 2) had been optimized in earlier experiments (Deimling *et al.*, 1990; Hörlein, 1991). The use of maltose has also resulted in greater repeatability of induction and regeneration rates among experiments over time (Daniel, 1993; Flehinghaus-Roux, 1994).

*Table 2.* Influence of sucrose and maltose in the induction medium on the induction (IR) and regeneration rate (RR) of DH5 and "Teku"

| Carbohydrate source | Concentration | DH5 | | TEKU | |
|---|---|---|---|---|---|
| | | IR (%) | RR (%) | IR (%) | RR (%) |
| Sucrose | 90 g $l^{-1}$ | 6.2 | 0.3 | 2.4 | 0 |
| Maltose | 120 g $l^{-1}$ | 25.0 | 3.0 | 5.3 | 0.1 |

The suitability of different carbohydrates for induction and regeneration of androgenic structures is dependent upon the species, although these preferences remain unclear. One hypothesis assumes that glucose could be toxic at higher concentrations (Hunter, 1987; Finnie *et al.*, 1989). An experiment was performed to measure the glucose content of liquid media 3, 6, 9, 12 and 15 days after initiation of rye anther culture that had been supplemented with either sucrose or maltose (Deimling *et al.*, 1992). A four-fold increased glucose concentration in the sucrose medium was found after three days of culture. Carbohydrate hydrolysis was much slower in maltose supplemented media. This may lead to a more favorable supply of glucose for androgenic development. Rye produced calli or embryos only on the maltose supplemented medium (Deimling *et al.*, 1992).

Plant growth regulators have generally been used for anther culture of rye. We have employed a combination of 2 mg $l^{-1}$ 2,4–dichlorophenoxyacetic acid (2,4–D) and 0.5 mg $l^{-1}$ kinetin (kin). Daniel (1993) used a combination of picloram, IAA, and BA. Different combinations of 2,4–D, benzyladenine (BA) and picloram were tested by Hörlein (1991), resulting in a slight superiority of the picloram supplemented medium, but significant differences were not obtained. Thomas *et al.* (1975) successfully regenerated plants from anther culture of rye using 2,4–D as the only growth regulator in their induction media. Several experiments have been performed in our laboratory with two different basal media (N6 and Olsen) in combination with various plant growth regulators (results not shown) with no consistent trends.

The gelling agent is another factor that can influence anther culture response. Since Wernicke & Kohlenbach (1976) demonstrated inhibitory effects of agar, many media have been preferentially solidified with agarose or gelrite. For rye anther culture, gelrite and agarose appear to be interchangeable, with induction and regeneration rates on gelrite solidified media slightly but not always significantly higher than those on agarose media (Fig. 3). Although this increase was not found for all genotypes tested, we recommend the use of gelrite generally in rye anther culture. Different concentrations of solidifying agents (agarose at 2.5 or 4.0 g $l^{-1}$; gelrite at 1.25 and 2.5 g $l^{-1}$) did not influence the response rates significantly (Flehinghaus *et al.*, 1991). Daniel (1993) regenerated plants on medium with 3.2 g $l^{-1}$ gelrite.

The choice of sterilization method may affect the anther culture response; autoclaving has the disadvantage that the exact composition of the medium

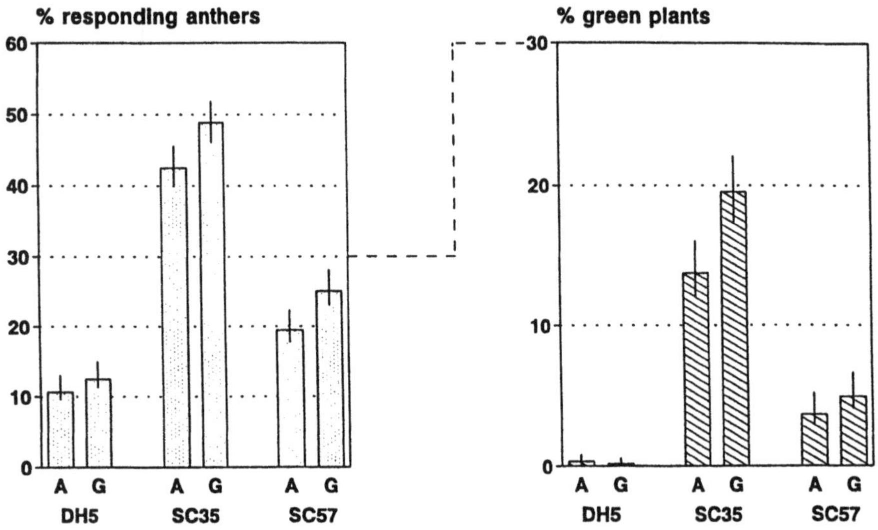


*Figure 3.* Influence of the gelling agents agarose (A) and gelrite (G) on the induction and regeneration rates of rye genotypes DH5, SC35 (DH3 × DH5), and SC57 (DH5 × DH7; vertical bars indicate the 95% confidence interval).

is unknown after the process because some compounds (especially carbohydrates, vitamins, and plant growth regulators) are thermally unstable (Pierik, 1987) and even the pH of the medium can be affected (Owen *et al.*, 1991). Filter-sterilization of media has been recommended by several authors, but this process is time consuming and costly, especially when anther culture is applied in commercial plant breeding and becomes a large scale process. A comparison of the two sterilization methods demonstrated a generally detrimental effect of autoclaving (Fig. 4). Autoclaving significantly reduced the induction rate of two genotypes (Y5, CF1). In addition, a dramatic decrease in the regeneration rate of SC35 was observed due to autoclaving. These negative effects lead to the conclusion that autoclaved medium should only be used for haploid rye production with highly responsive genotypes and low labour capacity. If possible, filter-sterilization should be preferred.

The culture conditions during induction and regeneration of rye anther cultures have not received much attention. The most commonly applied incubation conditions during induction have been 24–28°C in the dark (Wenzel & Thomas, 1974; Flehinghaus *et al.*, 1991; Daniel, 1993). Incubation under the same temperature regime but under low light intensity was preferred by Orlikowska (1977), Sharma *et al.* (1982), and Milewska-Pawliczuk (1987). More detailed investigations on culture conditions were carried out by Lorenz (1989). According to her results, the application of light did not increase the number of responsive anthers, but increased the callus viability compared to cultures incubated in the dark. Investigations on the light spec-

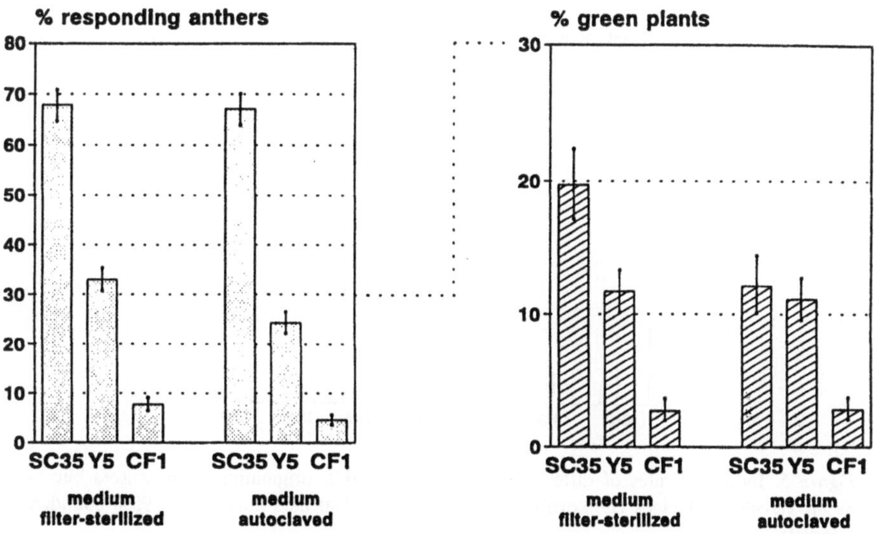


*Figure 4.* Influence of autoclaving and filter-sterilization of the induction medium on the induction and regeneration rates of rye genotypes SC35 (DH3 × DH5), double-cross Y5 (CF1 × SC35) and CF1 (L285-F × L283-R; vertical bars indicate the 95% confidence interval).

trum (white, red, and blue) and on incubation temperature (20 vs. 25°C) again only positively affected callus viability and not anther response. The influence of the light and temperature parameters during induction on the subsequent regeneration frequency could not be studied because the development of calli into plants failed in these experiments.

### 2.2.4. *Genotype*

In publications on rye anther culture (Table 1), induction of macroscopic structures has been achieved from a diverse genetic array of donor plants. In contrast, the regeneration of anther-derived plants seems to be highly dependent on the species. In cases where true *S. cereale* genotypes were tested in anther culture, regeneration of green plants was either impossible (Orlikowska, 1977; Sharma *et al.*, 1982; Milewska-Pawliczuk, 1987; Lorenz, 1989) or rare (Wenzel *et al.*, 1976, 1977; Hoffmann & Wenzel, 1981; Flehinghaus *et al.*, 1991; Hörlein, 1991).

Zenkteler & Misiura (1974) tested *S. montanum* in anther culture: under optimal conditions, multicellular pollen grains were found in about 1% of the inoculated anthers. The globular embryos that subsequently developed could not be regenerated into plants. Introgression of *S. vavilovii* into *S. cereale*, however, has increased the anther culture response. Thomas *et al.* (1975) plated more than 92 000 anthers of crosses between *S. cereale* cv. Heines Hellkorn and *S. vavilovii* (Kuckuck, 1976a,b) from which they obtained 149 macroscopic structures and eight green plants. In subsequent

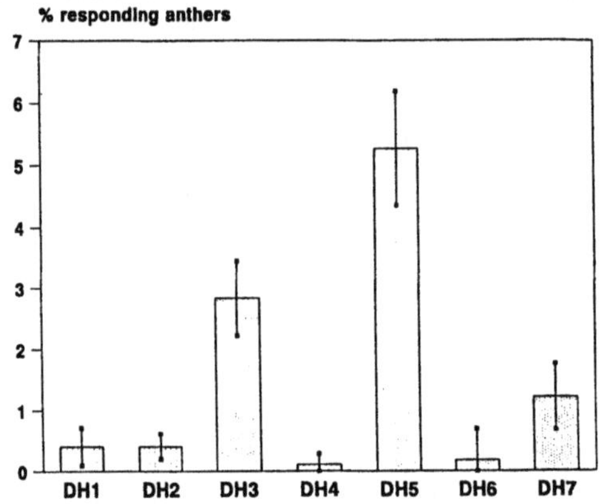


*Figure 5.* Induction rates of different anther-derived DHL originating from a cross between *Secale cereale* cv. Heines Hellkorn and *S. vavilovii* (vertical bars indicate the 95% confidence interval).

experiments, they increased regeneration rates in selected crosses partially containing the exotic germplasm (Wenzel *et al.*, 1977). Hörlein (1991) tested seven anther-derived DHL from these crosses (kindly provided by Prof. G. Wenzel, Grünbach, Germany) for anther culture ability. Significant differences in the induction rate were obtained ranging from 0.1% for DH4 to 5.3% for DH5 (Fig. 5). We then intermated the most responsive of the DHL (giving SC35 and SC57). Compared to DH5, a remarkable increase in anther culture response was observed in the single crosses (see Fig. 3). Up to 49% responsive anthers and 20% green plants were observed under optimal conditions using the superior single cross SC35. In addition to the beneficial effects of the *S. vavilovii* background, the increased vigour of the heterozygous $F_1$ material may have lead to further increased anther culture ability. Heterotic effects in anther culture were also found by Barloy *et al.* (1989) for maize. Selection for recombinant microspores carrying genes for androgenic capacity from both DH parents may also explain this phenomenon. Due to its extremely good anther culture ability (70–80% induction rate and around 20–25% regeneration rate under optimal conditions), the single cross SC35 has become a standard genotype in many laboratories and plant breeding companies working on rye anther culture. Using SC35, Daniel (1993) obtained a lower induction frequency of 31% but a higher regeneration rate of 29.8%.

To study the superiority of SC35 in anther culture, we compared the anther culture ability of its parental lines, DH3 and DH5. The induction rate of DH5 was significantly greater than that of DH3 (Fig. 6). In contrast, although the number of responding anthers was considerably lower for DH3, its regeneration rate was higher than that of DH5. Each parent seemed to

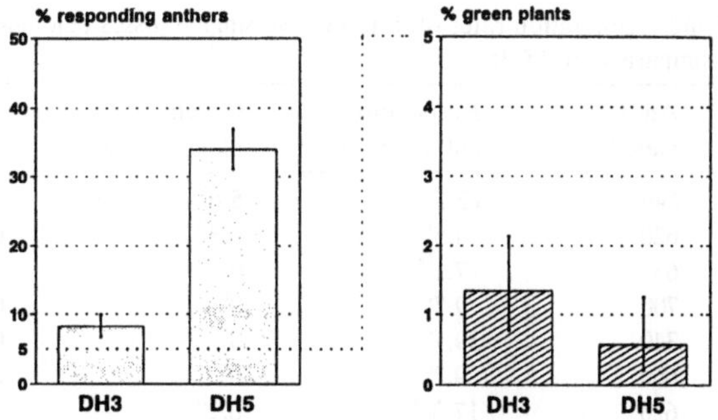


*Figure 6.* Induction rates (% responding anthers) and regeneration rates (% green plants) of rye genotypes DH3 und DH5 (vertical bars indicate the 95% onfidence interval).

express to a different degree the separate components that comprise the trait, anther culture ability. Segregation of this trait into distinct components has also been observed in other species (Deaton *et al.*, 1987; Singsit & Veilleux, 1988; Agache *et al.*, 1989).

Due to the poor regeneration capacity of *S. cereale*, the incorporation of the anther culture technique into practical rye breeding has been limited until recently (Lind *et al.*, 1985; Wenzel & Foroughi-Wehr, 1990). Wenzel & Thomas (1974) first reported the sporadic regeneration of green plantlets among genotypes of *S. cereale*. The successful development of DHL of *S. cereale* has been laborious and infrequent. A few DHL have been established from the tetraploid ($2n = 4x = 28$) cultivars "Tero" (Wenzel *et al.*, 1977) and "Teku" (Hörlein, 1991; Flehinghaus *et al.*, 1991). In diploid ($2n = 2x = 14$) material, Wenzel *et al.* (1976) regenerated plants from the cross "Pekuro" × "Moscow Dwarf" and later from L281 × L282 (Wenzel *et al.*, 1985). Daniel (1993) reported a higher regeneration rate of 3.3% from the line, L201. Recently, the successful regeneration of DHL from "Danko" and "Carokurz" has been published (Flehinghaus-Roux, 1994).

In order to detect higher anther culture ability in true *S. cereale* material, we examined different single crosses within the distinct genepools "Petkus" and "Carsten" for their anther culture response in comparison with the partially unadapted standard, SC35. Again, the development of androgenic structures was obtained from all *S. cereale* genotypes except for one $F_1$ of the "Petkus" background (Flehinghaus-Roux *et al.*, 1995). The induction rate in the "Petkus" genepool was rather low and did not exceed 6.8%. Regeneration of green plants was not possible from the "Petkus" genetic background. However, the situation was different in the "Carsten" genepool. Of three "Carsten" $F_1$ tested, we identified a single cross, L285-F × L283-R, that regenerated five green plants from 857 anthers plated, although this cross showed the lowest induction rate (3.7%). In a second step, we tested the parents (L285–F and L283–R) of this promising single cross in different

*Table 3.* Induction and regeneration rates of different rye single-crosses belonging to the "Carsten" genepool in comparison to SC35

| Genotype | Anthers plated (no.) | Responding anthers (%) | Green plantlets (%) | Albino plantlets (%) |
|---|---|---|---|---|
| SC35 | 540 | 63.5[a] | 5.6[a] | 0[a] |
| L285-F × L283-R | 620 | 4.5[b] | 0.3[b] | 0[a] |
| L283-R × L285-R | 661 | 7.1[bc] | 0.6[b] | 0[a] |
| L283-R × L287-R | 700 | 9.7[cd] | 0[b] | 0[a] |
| L283-R × L289-R | 340 | 39.7[e] | 0[b] | 0.9[a] |
| L283-R × L290-R | 660 | 9.7[c] | 0[b] | 0[a] |
| L285-R × L287-R | 620 | 17.3[cd] | 0.2[b] | 0.3[a] |
| L285-R × L289-R | 600 | 26.8[de] | 1.2[b] | 0[a] |
| L285-R × L290-R | 680 | 17.7[e] | 0.5[b] | 0.2[a] |

[a,b,c] Means within columns followed by the same letter belong to overlapping confidence intervals ($P = 0.95$).

combinations with other lines from the "Carsten" genepool (Table 3). Although all crosses tested showed induction ability (ranging from 4.5 to 39.7%), only combinations with L285 were able to regenerate green plants. Thus, this line seems to carry genes for regeneration ability (Flehinghaus-Roux *et al.*, 1995). The high androgenic capacity of SC35 could not be reached by any other genotype tested.

The data obtained clearly show that anther culture ability is genetically controlled. The double cross, Y5, that possessed only 25% *S. vavilovii* background, showed approximately intermediate induction and regeneration rates compared to its single cross parents, CF1 and SC35 (Fig. 4). Anther culture response was transferred from the high responding genotype, SC35, that had 50% *S. vavilovii* background, into the low responding true *S. cereale* cross CF1. The regeneration rates obtained were considerably higher than those reported by Wenzel *et al.* (1976) who tested double crosses containing a similar percentage of the exotic germplasm.

### 2.3. *Regeneration and maintenance of green plants*

#### 2.3.1. *Development of androgenic structures and plant growth*

Regeneration of anther-derived calli and embryogenic structures of rye showed best results on a modified N6 medium (Chu, 1978) containing only kin ($2\,mg\,l^{-1}$) as growth regulator (Flehinghaus *et al.*, 1991). In contrast, Wenzel *et al.* (1977) and Daniel (1993) regenerated similar structures on a medium free of growth regulators. Hörlein (1991) tested 11 regeneration media differing by growth regulator supplements. She found low auxin concentrations necessary for further callus development towards green plantlets. On the growth regulator-free variant, callus development ceased in her investigations. A definitive conclusion on the suitability of plant growth regulators in the regeneration process cannot yet be made.

Due to limited regeneration capacity in rye, the number of experiments concerning regeneration media and optimal conditions is rather low. With our protocol, a further transfer of the green plantlets from regeneration medium containing kin to growth regulator free medium is necessary to obtain sufficient rooting of the plants. Culture conditions during the regeneration process do not seem to be critical. Slow adaptation of green plantlets to high light intensity was beneficial (Flehinghaus *et al.*, 1991). The optimal daylength and temperature of 14–16 h and 22–27°C, respectively, have been recommended by several authors.

2.3.2. *Vernalization and* in vitro *multiplication*

After haploid rye plantlets have developed sufficient roots, those derived from winter cultivars need to be vernalized. Two methods have been successful. Daniel (1993) transferred plants at the 2–3 leaf stage from the culture tube into peat soil for hardening (20°C, high humidity). After 14 days, he vernalized them for ten weeks at 4°C. Thereafter, he grew the plants to maturity under standard greenhouse conditions. To save labour and keep plant losses during vernalization low, we followed a different strategy. Regenerants were vernalized *in vitro* by transferring them directly from 27°C to the vernalization chamber at 2°C for 10 weeks under low light conditions. Thereafter, the plantlets, still *in vitro*, were transferred for adaptation to 14°C and intermediate light conditions for 14 days before they were potted in soil, kept under high humidity for a few days and grown to maturity under normal greenhouse conditions. Problems with fungal infection can be avoided by this aseptic vernalization. In addition, synchronization of regenerants is facilitated because the vernalization period can be prolonged quite easily. This protocol enabled us to group plantlets into larger lots because *in vitro* storage could be extended for more than 12 months without changing medium. It is unknown whether this method has an influence on ploidy, flowering or fertility of the regenerants.

Multiplication of the regenerated haploids *in vitro* prior to vernalization might be advisable for several reasons (Hörlein, 1991). Firstly, the probability that a haploid/doubled haploid genotype will survive until flowering can be increased. Secondly, seed-set after self-pollination can be sufficiently increased to allow a field trial with $R_1$ generation seed even in cases where flowering may be weak. If chromosome doubling is necessary, more copies of each clone enhance the chance of success.

Substances that are known to induce or increase tillering have been applied to rye cultures to achieve *in vitro* multiplication. Media supplemented with BA have shown a positive influence on budding (Ryschka & Leike, 1986). Ethylene or substances releasing ethylene have been shown to promote tillering in rye cultures (Köhler & Nestrowicz, 1989). In our laboratory, BA application for four weeks at concentrations ranging from 1.25 to 20.0 mg $l^{-1}$ did not result in any significant difference compared to the untreated control. However, the average number of tillers per plant doubled for five genotypes after a four week incubation at 22°C in media containing 0.07 ml $l^{-1}$ of the ethylene releasing compound, "Cerone" (Union-Carbide) (unpublished data). Using the same concentration, incubation at 15°C also gave a significant but smaller increase that was likely due to the slower release of ethylene at cooler temperatures. After transferring "Cerone"-treated plants to basal medium, the process of tillering continued. A shorter application may have been sufficient for tillering; longer durations were ineffective.

### 2.3.3. *Ploidy of regenerants and fertility*

Occurrence of spontaneous doubling of haploids to doubled haploids has been frequently reported for rye. In early work, Wenzel & Thomas (1974) stated that most of their anther-derived plants were at the 2*x* level; later they observed 100% spontaneous chromosome doubling (Thomas *et al.*, 1975; Wenzel *et al.*, 1976). The majority of plants regenerated by Hörlein (1991) was spontaneously doubled. These data were obtained from only a small number of regenerants. From their experience over a longer period of time, Wenzel & Foroughi-Wehr (1984) estimated the spontaneous doubling rate for rye at approximately 70%. Our own data confirmed this: with SC35 and CF1, 66 and 75% of the regenerants examined had spontaneously doubled (Flehinghaus-Roux, 1994). The rate of haploids was 17 and 8% for SC35 and CF1, respectively. The remaining plants were either aneuploid, mixoploid, triploid or tetraploid.

Artificial doubling of rye haploids has also been attempted but without great success. By treating regenerated haploids for 5 h with 0.5% aqueous colchicine, Wenzel *et al.* (1977) obtained only 20% of regenerants with altered chromosome number; unfortunately, 50% of the plants did not survive the treatment. Following the doubling protocol of Jensen (1974) with various species, Wenzel & Foroughi-Wehr (1984) successfully doubled the chromosome number of wheat, rice, and barley but found for rye that the procedure often had to be repeated to generate any doubled haploids.

Variable fertility has been reported among doubled haploid plants regenerated from anther culture of rye. Hoffmann & Wenzel (1981) observed seed set after selfing on only 21 of 40 (53%) regenerants that developed spikes. Hörlein (1991) obtained $R_1$ seeds from 13 of 27 (48%) doubled haploids. Of 70 regenerated plants of SC35 that were self-pollinated, 46 (66%) set seed whereas 23 of 37 CF1 plants (62%) could generate DHL after self-pollination under greenhouse conditions. Regenerants ($R_0$-plants) that were planted directly in the field to natural vernalization and growth conditions exhibited approximately 30% seed set after selfing. The reasons for the poor fertility are unknown.

### 2.3.4. *Field performance of DHL*

Although DHL of rye have only rarely been available, there have been some promising data on their agronomic performance. For all characters (100 kernel weight, plant height, ear length, and alkylresorcinol content) examined in a field trial on a small number of DHL conducted by Hoffmann & Wenzel (1981), the means of the anther-derived lines significantly differed from those of the inbred controls in both directions. Friedt *et al.* (1983) tested the value of DHL derived from *S. cereale* × *S. vavilovii* crosses. They found variation among DHL for different morphological and agronomic characters, such as seed set, 1000–grain-weight, and self-fertility. In topcross tests, some DHL exhibited positive general combining ability. Lind *et al.* (1985) and Wenzel *et al.* (1985) stated that the low general index (composite of different agronomic traits weighted according to their importance) of these DHL was mainly caused by the negative genetic contribution of the unadapted parent, *S. vavilovii*.

Recently, we evaluated 150 DHL, including, for the first time, a large number of true *S. cereale* doubled haploids, in a preliminary field study. Distinct morphological differences among DHL originating from a single cross could be observed. Variation within DHL did not occur. Regular field trials and breeding behavior of the best DHL will be conducted.

### 2.4. *Albinism*

In anther culture of monocots, the occurrence of albino plants has often been observed. In wheat (Tuveson *et al.*, 1989) and barley (Larsen *et al.*, 1991), albinism seems to be genetically controlled. Day & Ellis (1984) stated that albino plants regenerated from anther culture show deletions in the chloroplast genome. The deficiency of plastid DNA in albinos and the absence of fraction I proteins were reported by Wang *et al.* (1978) and Sun *et al.* (1979). The development of albinos may also be due to mutations or the expression of recessive genes (Clapham, 1977).

Anther culture of rye has also been characterized by a high frequency of albinos. Early experiments with crosses between *S. cereale* and *S. vavilovii* revealed albino rates ranging from 70.4% to 89.7% (Thomas *et al.*, 1975; Wenzel *et al.*, 1976, 1977). More recent data indicate a clear genotypic influence on the albino frequency (Table 4). Except for line L201 (Daniel, 1993), a higher rate of albino plants has been regenerated in *S. cereale* compared to genotypes partially possessing *S. vavilovii* genetic background. In some instances ("Carokurz" or germplasm from the "Petkus" genepool), 100% albino plants were regenerated. For "Carsten" single-crosses containing L283, L289, and L290 as one parental line, many albinos were also regenerated (Table 3). The genotypic influence is especially evident in a comparison of the crosses L283-R × L289-R and L285-R × L289-R. All regenerants of the first cross were albinos, whereas the latter produced 80% green plants. L285 might contain factors positively influencing the ratio between green and albino plants. Large differences in albino frequency of various $F_1$ have also been reported by Wenzel *et al.* (1977).

Although the genotypic effect is evident, other factors prior to or during

*Table 4.* Summary of investigations on albinism rates after anther culture of different rye genotypes originating from *Secale cereale* × *S. vavilovii* crosses or from true *S. cereale*

| Authors | Year | *Secale*-background | Genotype | Regeneration | |
|---|---|---|---|---|---|
| | | | | Total no. | % albinos |
| Lind *et al.* | 1985 | *S.c.* × *S.v.* | | 173 | 36.8 |
| | | *S.c.* | L281 × L282 | 8 | 75.0 |
| Daniel | 1993 | *S.c.* × *S.v.* | SC35 | 322 | 26.4 |
| | | *S.c.* | "Carokurz" | 216 | 100.0 |
| | | *S.c.* | L201 | 56 | 0 |
| Flehinghaus-Roux *et al.* | 1995 | *S.c.* × *S.v.* | SC35 | 123 | 0 |
| | | *S.c.* | 6 $F_1$ ("Petkus") | 3 | 100.0 |
| | | *S.c.* | 10 $F_1$ ("Carsten") | 36 | 63.6 |

*in vitro* culture can also affect the frequency of albinos regenerated. Change of osmolality or maltose concentration led to a shift from a high albino rate towards an increased number of green plants in wheat (Zhou *et al.*, 1991). This observation was confirmed by our own experiments with different maltose concentrations on rye anther culture (data not shown). However, in both cases, the total regeneration rate decreased. By comparing the albino rates (25 vs. 0%) obtained with SC35 by Daniel (1993) and Flehinghaus-Roux *et al.* (1995), it is evident that factors other than genotype influence the frequency of albinism (Table 4). Wenzel & Thomas (1974) often observed that both green and albino plantlets could arise from the same callus. They attributed this result to either chimerism or the influence of physiological factors.

### 3. Isolated microspore culture

Due to the fact that rye is one of the most recalcitrant graminaceous species even in anther culture, reports of isolated microspore culture in rye have been rare. Wenzel *et al.* (1975) isolated microspores either directly from the spikes or from precultured anthers but were unable to regenerate plants in either case. However, by washing, filtering, and centrifuging the crude microspore preparation, they were able to obtain a purified fraction containing microspores predominantly at the late uninucleate stage. They also optimized conditions for maintaining microspore viability in culture by increasing mannitol concentration in the medium; however, the percent dividing microspores decreased if the mannitol concentration exceeded 0.5 M.

Deimling *et al.* (1994) recently reported the successful regeneration of plants from isolated microspores of SC35, the genotype that had been especially developed for enhanced anther culture response (Table 3). Isolation of microspores using a "Waring" blender gave better results than a mechanical isolation procedure with sieves. A one-week preculture of anthers in 7% mannitol at 14°C was an essential prerequisite to callus induction and subsequent regeneration. An additional cold-treatment (14°C, 6 d, dark) to microspores directly after isolation similar to that applied in anther culture, however, led to a dramatic drop in microspore viability. Untreated control microspores produced more green plants than those that were cold-treated for 6 days. In contrast to anther culture, where the optimal maltose concentration was 120 g $l^{-1}$, 60 g $l^{-1}$ was best in isolated microspore culture. The highest frequency of callus induction was observed at the highest microspore density ($1.5 \times 10^5$ $ml^{-1}$) tested. Once callus induction had been obtained with SC35, regeneration of plants was routine (Deimling, Röber, and Geiger, in preparation).

Compared to anther culture, where the albino rate was negligibly low with SC35 (Flehinghaus-Roux, 1994), an albino frequency of 43% was observed among regenerants of isolated microspores from the same genotype. Regeneration rates of approximately 10% green plants (isolating microspores from 100 anthers) were observed. Simultaneous experiments were undertaken with our best responding *S. cereale* genotype, CF1. Although callus could be frequently induced, regeneration of plants was not possible from this material.

## 4. Conclusions and prospects

Rye anther and microspore culture techniques have been considerably improved over the past 20 years. Methods and protocols have been developed that allow the reproducible production of androgenic rye plants by anther culture. Doubled haploid plants can be regenerated at predictable rates not only from semi-primitive genotypes of *S. vavilovii* origin but also from selections of *S. cereale* with more promising breeding potential. However, the overall regeneration rates are still too low for the general production of DHL as parental lines for hybrid breeding. Anther-derived DHL can be used to address specific breeding problems such as the derivation of homozygous forms of disease resistance or fertility restorers.

Because the anther culture procedure has recently been refined for application to cultivated forms of rye, the technique is now in practice by commercial plant breeding companies. Nevertheless, its practical use in current breeding programs is still at the beginning. However, the anther culture process itself appears to exert an effective selection for genes that favour androgenesis. In maize, a single cycle of selection resulted in a greater than six-fold increase in anther culture responsiveness (Petolino *et al.*, 1988). As a consequence, routine application of the anther culture technique in applied rye breeding should automatically increase the "anther culture ability" of breeding material.

Although it would be desirable to remove the genotypic limitation of

*Table 5.* Sample timetable of one cycle of rye anther culture from sowing donor plant material until harvest of $R_1$ seeds after self-pollination of regenerants

| Month/Year | Activity | Duration |
|---|---|---|
| October 1991 | sowing of donor plant material, germination and growth of young plantlets | 1 week |
| | vernalization period | 10 weeks |
| | adaptation period | 2 weeks |
| | growth of donor plants (greenhouse) | 8 weeks |
| March 1992 | harvest of donor spikes, cold treatment of spikes | 1 week |
| | anther plating, post plating temperature treatment | 1 week |
| | induction period | 12 weeks |
| June 1992 | transfer of callus, regeneration period | 4 weeks |
| July 1992 | vernalization of regenerants | 10 weeks |
| | adaptation period | 2 weeks |
| October 1992 | transfer of regenerants to soil, plant growth and self-pollination (greenhouse) | 22 weeks |
| March 1993 | harvest of $R_1$ seeds | |

anther culture, a more realistic goal is to transfer anther culture ability via genetic means. In backcross programs, it is possible to transfer genes for anther culture ability into germplasm lacking this trait. As donor, genotypes partially originating from *S. vavilovii* could be used. Due to the negative characters of this exotic germplasm (lodging, brittle spikes, and low grain yield), several backcross generations would be needed. A stepwise combination of genes for anther culture ability from different *S. cereale* donors may be preferred. However, this would be time consuming, because a test for anther culture response would be required in each backcross generation. The incorporation of genes for anther culture ability into advanced breeding material would be faster and more effective if such genes could be associated with closely linked markers. Using RFLP markers, the identification of such genes has been possible in maize (Cowen *et al.*, 1992). Similar success could be expected through marker assisted selection in rye.

The time saving advantage due to the introduction of DHL in plant breeding has often been cited. This argument has to be analyzed more critically if DHL production takes longer than one year (Strahwald, 1988). In our experience, a complete anther culture cycle from planting seed of donor plants until harvest of $R_1$ seeds requires approximately 18 months for winter rye (Table 5). When $R_1$ seed formation on $R_0$ plants that have been derived from anther culture is insufficient, a second generation ($R_2$) may be required for evaluation of the DHL. Field performance of androgenically derived rye DHL can be evaluated two and a half years after donor plant material was sown, at the earliest. Thereafter, time is needed for testcrossing and selection. In conventional plant breeding, selection is performed during early inbred generations leading to an enormous reduction of the number of lines that need further testing. Using DHL, tests on line performance and combining ability are required with a large number of genotypes.

In conclusion, rye anther culture has reached a standard where methodical improvements of the protocol will not likely lead to drastic breakthroughs. Further advances in derivation and utilization of haploid rye are more likely to be achieved via genetic means. To date, anther culture and the subsequent production of DHL can be successfully used to solve special breeding problems. Whether DHL can be effectively incorporated in hybrid rye breeding will be answered during the next few years through the ongoing investigations of various scientific groups and breeding companies.

## 5. Acknowledgements

The authors thank Anke Pelzer for her excellent technical assistance over the past five years. Thank you to Julia Magnussen and Frank Röber for their help preparing this review and to Prof. H.H. Geiger and Dr. J.H. Werner for their critical and helpful comments on the manuscript.

## 6. References

Agache, S., B. Bachelier, J. De Buyser, Y. Henry & J. Snape (1989) Genetic analysis of anther culture response in wheat using aneuploid, chromosome substitution and translocation lines. Theor. Appl. Genet. 77: 7–11.

Anonymous (1976) A sharp increase of the frequency of pollen plant induction in wheat with potato medium. Acta Genet. Sin. 3: 25–31.

Barloy, D., L. Denis & M. Beckert (1989) Comparison of the aptitude for anther culture in some androgenetic doubled haploid maize lines. Maydica 34: 303–308.

Büter, B., J.E. Schmid & P. Stamp (1991) Effects of L-proline and post-plating temperature treatment on maize (*Zea mays* L.) anther culture. Plant Cell Rep. 10: 325–328.

Chu, C.C. (1978) The N6 medium and its application to anther culture of cereal crops. In: Proc. Symp. Plant Tissue Culture, pp. 43–50. Science Press, Beijing.

Clapham, D. (1977) Haploid induction in cereals. In: J. Reinert & Y.P.S. Bajaj (Eds.), Applied and Fundamental Aspects of Plant Cell Tissue and Organ Culture, pp. 279–298. Springer-Verlag, Berlin.

Cowen, N.M., C.D. Johnson, K. Armstrong, M. Miller, A. Woosley, S. Pescitelli, M. Skokut, S. Belmar & J.F. Petolino (1992) Mapping genes conditioning *in vitro* androgenesis in maize using RFLP analysis. Theor. Appl. Genet. 84: 720–724.

Daniel, G. (1993) Anther culture in rye: improved plant regeneration using modified MS-media. Plant Breed. 110: 259–261.

Day, A. & T.H.N. Ellis (1984) Chloroplast DNA deletions associated with wheat plants regenerated from pollen: possible basis for maternal inheritance of chloroplasts. Cell 39: 359–368.

Deaton, W.R., S.G. Metz, T.A. Armstrong & P.N. Mascia (1987) Genetic analysis of the anther-culture response of three spring wheat crosses. Theor. Appl. Genet. 74: 334–338.

Deimling, S., T. Flehinghaus, I. Schneider & H.H. Geiger (1990) Influence of post-plating temperature and carbohydrate source in maize and rye anther culture. In: Abstracts VIIIth International Congress on Plant Tissue Culture, Amsterdam, p. 195.

Deimling, S., T. Flehinghaus, M. Bohn & H.H. Geiger (1992) Sugary aspects of androgenesis – alterations of carbohydrates during haploid induction in gramineous species. In: Abstracts XIIIth EUCARPIA Congress, Angers, July 6–11, 1992, pp. 153–154.

Deimling, S., T. Flehinghaus-Roux, F. Röber, A. Schechert, S.R. Roux & H.H. Geiger (1994) Doubled haploid production – now reproducible in rye. In: Abstracts VIIIth International Congress of Plant Tissue and Cell Culture, Firenze, June 12–17, 1994, p. 95.

Dieu, P. & M. Beckert (1986) Further studies of androgenetic embryo production and plant regeneration from *in vitro* cultured anthers in maize (*Zea mays* L.). Maydica 31: 245–259.

Finnie, S.J., W. Powell & A.F. Dyer (1989) The effect of carbohydrate composition and concentration on anther culture response in barley (*Hordeum vulgare* L.). Plant Breed. 103: 110–118.

Flehinghaus, T., S. Deimling & H.H. Geiger (1991) Methodical improvements in rye anther culture. Plant Cell Rep. 10: 397–400.

Flehinghaus-Roux, T. (1994) Induktion und Regeneration androgenetischer Pflanzen bei Roggen. Ph.D. Thesis, University of Hohenheim, Stuttgart.

Flehinghaus-Roux, T., S. Deimling & H.H. Geiger (1995) Anther culture ability in *Secale cereale* L. Plant Breed. 114 (in press).

Foroughi-Wehr, B. & G. Mix (1979) *In vitro* response of *Hordeum vulgare* L. anthers cultured from plants grown under different environments. Environ. Exptl. Bot. 19: 303–309.

Friedt, W., V. Lind, H. Walther, B. Foroughi-Wehr, S. Züchner & G. Wenzel (1983) The value of inbred lines derived from *Secale cereale* × *S. vavilovii* via classical inbreeding and androgenetic haploids. Z. Pflanzenzüchtg. 91: 89–103.

Gaul, H., G. Mix, B. Foroughi-Wehr & M. Okamoto (1976) Pollen grain development of *Hordeum vulgare*. Z. Pflanzenzüchtg. 76: 77–80.

Geiger, H.H. (1972) Wiederherstellung der Pollenfertilität in cytoplasmatisch männlich sterilem Roggen. Theor. Appl. Genet. 42: 32–33.

Geiger, H.H. (1982) Breeding methods in diploid rye (*Secale cereale* L.). Tag.- Ber., Akad. Landwirtsch.-Wiss. DDR, Berlin 198: 305–332.

Geiger, H.H. (1985) Einsatzmöglichkeiten biotechnologischer Verfahren in der Resistenz-züchtung. In: Biotechnologie in der Agrarwissenschaft, pp. 57–74. University of Saarland, Saarbrücken.

Geiger, H.H. (1990) Wege, Fortschritte und Aussichten in der Hybridzüchtung. In: G. Haug, G. Schuhmann & G. Fischbeck (Eds.), Pflanzenproduktion im Wandel, pp. 41–72. VCH Verlagsges., Weinheim.

Geiger, H.H. & F.W. Schnell (1970) Cytoplasmic male sterility in rye (*Secale cereale* L.). Crop Sci. 10: 590–593.

Hoffmann, F. & G. Wenzel (1981) Selfcompatibility of microspore-derived double-haploid rye lines and single grain selection for alkylresorcinol content. Theor. Appl. Genet. 60: 129–133.

Hörlein, A.J. (1991) Methodische Untersuchungen zur Antherenkultur bei Roggen. Ph.D. Thesis, University of Hohenheim, Stuttgart.

Hunter, C.P. (1987) European Patent Application by Shell International Research Maatschappij B.V., No. 87200773.7.

Jähne, A., D. Becker, R. Brettschneider & H. Lörz (1994) Regeneration of transgenic, microspore-derived, fertile barley. Theor. Appl. Genet. 89: 525–533.

Jensen, C.J. (1974) Chromosome doubling techniques in haploids. In: K.J. Kasha (Ed.), Haploids in Higher Plants: Advances and Potential, pp. 153–190. University of Guelph, Guelph, Ontario.

Köhler, H. & N. Nestrowicz (1989) Verfahren zur Erzeugung und Depothaltung genetisch identischer Nachkommen von Getreidegenotypen. Patentschrift der Deutschen Demokratischen Republik, Wirtschaftspatent DD 264 938 Al.

Kuckuck, H. (1976a) Die Verwendung der Wildart *Secale vavilovii* Grossh. in der Roggenzüchtung. Züchtertagung Gumpenstein Bericht, pp. 144–159.

Kuckuck, H. (1976b) Spontane Kreuzungen zwischen *Secale vavilovii* Grossh. ♀ und *Secale cereale* L. ♂. Z. Pflanzenzüchtg. 77: 1–9.

Larsen, E.T., I.K.D. Tuveson & S.B. Anderson (1991) Nuclear genes affecting percentage of green plants in barley (*Hordeum vulgare* L.) anther culture. Theor. Appl. Genet. 82: 417–420.

Laurie, D.A., L.S. O'Donoughue & M.D. Bennett (1990) Wheat × maize and other wide sexual hybrids: their potential for genetic manipulation and crop improvement. In: J.P. Gustafson (Ed.), Gene Manipulation in Plant Improvement. II, pp. 95–126. Plenum Press, New York.

Lind, V., B. Foroughi-Wehr & G. Wenzel (1985) The variation of mildew resistance within and between population varieties of rye and prospects for the production of inbred lines via anther culture. In: Proceedings of the EUCARPIA meeting of the cereal section on rye, June 11–13th 1985, Svalöv, pp. 391–408.

Lorenz, I. (1989) Ergebnisse zur Induktion der androgenetischen Entwicklung in Winter-roggenantheren. Arch. Züchtungsforsch. Berlin 19(6): 415–420.

Marsolais, A.A., G. Séguin-Swartz & K.J. Kasha (1984) The influence of anther cold pretreatments and donor plant genotypes on *in vitro* androgenesis in wheat (*Triticum aestivum* L.). Plant Cell Tiss. Organ Cult. 3: 69–79.

Melchinger, A.E. (1990) Use of molecular markers in breeding for oligogenic disease resistance. Plant Breed. 104: 1–19.

Milewska-Pawliczuk, E. (1987) Induction of androgenesis *in vitro* in various genetic forms of *Secale cereale*. Biol. Plant. 29: 295–298.

Murashige, T. & F. Skoog (1962) A revised medium for rapid growth and bioassays with tobacco tissue cultures. Physiol. Plant. 15: 473–497.

Olsen, F.L. (1987) Induction of microspore embryogenesis in cultured anthers of *Hordeum vulgare*. The effects of ammonium nitrate, glutamine and asparagine as nitrogen sources. Carlsberg Res. Commun. 52: 393–404.

Orlikowska, T. (1977) Induction of androgenesis *in vitro* in *Secale cereale* and *Triticale*. Genet. Polonica 18: 51–59.

Owen, H.R., D. Wengerd & A.R. Miller (1991) Culture medium pH is influenced by basal medium, carbohydrate source, gelling agent, activated charcoal, and medium storage method. Plant Cell Rep. 10: 583–586.

Petolino, J.F., A.M. Jones & S.A. Thompson (1988) Selection for increased anther culture response in maize. Theor. Appl. Genet. 76: 157–159.

Pierik, R.L.M. (1987) *In Vitro* Culture of Higher Plants. Martinus Nijhoff Publishers, Dordrecht.

Powell, W. (1988) The influence of genotype and temperature pre-treatment on anther culture response in barley (*Hordeum vulgare* L.). Plant Cell Tiss. Organ Cult. 12: 291–297.

Ryschka, S. & H. Leike (1986) Effects of BAP and Camposan on induction of lateral buds in cereals – condition for *in vitro* propagation. Arch. Züchtungsforsch. Berlin 16(2): 81–89.

Seibel W. (1988) Bedeutung als Kulturpflanze. In: W. Seibel & W. Steller (Eds.), Roggen: Anbau – Verarbeitung – Markt, pp. 15–23. Behr's Verlag, Hamburg.

Sharma, G.C., W.C. Wang & V.T. Sapra (1982) Effect of genotype, media and temperature pretreatment on callus initiation in triticale, wheat, and rye anther cultures. Cereal Res. Commun. 10: 143–150.

Singsit, C. & R.E. Veilleux (1988) Intra- and interspecific transmission of androgenetic competence in diploid potato species. Euphytica 43: 105–112.

Strahwald, J.F. (1988) Modellrechnungen zur Optimierung der Rekurrenten Selektion bei Gerste. Ph.D. Thesis, University of Hohenheim, Stuttgart.

Sun, C.S., S.C. Wu, C.C. Wang & C.C. Chu (1979) The deficiency of soluble proteins and plastid ribosomal RNA in the albino pollen plantlets of rice. Theor. Appl. Genet. 55: 193–197.

Thomas, E., F. Hoffmann & G. Wenzel (1975) Haploid plantlets from microspores of rye. Z. Pflanzenzüchtg. 75: 106–113.

Tuveson, I.K.D., S. Pedersen & S.B. Andersen (1989) Nuclear genes affecting albinism in wheat (*Triticum aestivum* L.) anther culture. Theor. Appl. Genet. 78: 879–883.

Wang, C.C., C.S. Sun, C.C. Chu & S.C. Wu (1978) Studies on the albino pollen plantlets of rice. In: Proceedings of Symposium on Plant Tissue Culture, pp. 149–160. Science Press, Peking.

Wenzel, G. & B. Foroughi-Wehr (1984) Anther culture of cereals and grasses. In: I.K. Vasil (Ed.), Cell Culture and Somatic Cell Genetics of Plants, Vol. 1, pp. 311–327. Academic Press Inc., New York.

Wenzel, G. & B. Foroughi-Wehr (1990) Haploidenkultur. Agrarspectrum 17: 85–98.

Wenzel, G. & E. Thomas (1974) Observations on the growth in culture of anthers of *Secale cereale*. Z. Pflanzenzüchtg. 72: 89–94.

Wenzel, G., F. Hoffmann & E. Thomas (1976) Heterozygous microspore-derived plants in rye. Theor. Appl. Genet. 48: 205–208.

Wenzel, G., F. Hoffmann & E. Thomas (1977) Increased induction and chromosome doubling of androgenetic haploid rye. Theor. Appl. Genet. 51: 81–86.

Wenzel, G., B. Foroughi-Wehr, W. Friedt & F. Köhler (1985) Anther culture in crop plants. In: Biotechnology in International Agricultural Research, Proceedings, Manila, April 23–27, 1984, pp. 65–73.

Wenzel, G., F. Hoffmann, I. Potrykus & E. Thomas (1975) The separation of viable rye microspores from mixed populations and their development in culture. Mol. Gen. Genet. 138: 293–297.

Wernicke, W. & H.W. Kohlenbach (1976) Investigations on liquid culture medium as a means of anther culture in *Nicotiana*. Z. Pflanzenphysiol. 79: 189–198.

Zenkteler, M. & E. Misiura (1974) Induction of androgenic embryos from cultured anthers of *Hordeum*, *Secale*, and *Festuca*. Biochem. Physiol. Pflanzen 165: 337–340.

Zhou, H., Y. Zheng & C.F. Konzak (1991) Osmotic potential of media affecting green plant percentage in wheat anther culture. Plant Cell Rep. 10: 63–66.

# 11. Oat haploids from anther culture and from wide hybridizations

HOWARD W. RINES, OSCAR RIERA-LIZARAZU, VICTOR M. NUNEZ, DOUGLAS W. DAVIS and RONALD L. PHILLIPS

**Contents**



## 1. Introduction

The common cultivated oat (*Avena sativa* L.) is one of the world's important cereal crops, with grain production currently at about 33 million metric tons per year (U.S. Dept. Agric. Statistics Service, 1994). Oat plant growth is favored by cool climates with the primary production occurring in the cooler temperate regions of the northern and southern hemispheres. Although there are fall-sown cultivars, spring-sown cultivars account for most of the oat grain production because of a general lack of winter hardiness in oat compared to other small grains. The proportion of land planted to oat has in general declined over the past 30 years, giving way to higher value small grains or oilseed crops; however, oat is still often grown for one or more of its attributes including high protein quality and content, good forage and bedding for livestock, a companion crop for forage legume establishment, adaptation to particular climatic zones, and as an integral part of a crop rotation program. Breeders' concerns in addition to increased productivity of high quality grain and forage are primarily focused on disease resistance, particularly toward the fungal pathogens including rusts, smuts, and mildews

*S.M. Jain, S.K. Sopory & R.E. Veilleux (eds.)*, In Vitro *Haploid Production in Higher Plants, Vol. 4*, 205–221.

and toward various viruses including barley yellow dwarf virus. There is also recent interest in developing specialty oat cultivars with modified grain composition for enhanced human or livestock nutrition or for possible specialty industrial uses.

The common cultivated oat is similar to bread wheat (*Triticum aestivum* L.) in being an allohexaploid with a chromosome complement of $2n = 6x = 42$. Haploids of hexaploid oat thus have a 21 chromosome complement consisting of three genomic sets of seven chromosomes and could also be referred to as "polyhaploids," as has been done in wheat by some authors (e.g., Mujeeb-Kazi & Riera-Lizarazu, 1996). The three genomic chromosome sets in hexaploid oat presumably each trace to a diploid ancestral donor and are designated the A, C, and D genomes. Although several diploid and tetraploid species of oat are known including possible A and C genome species, there is no extant species that fits the criteria for a direct genome donor in hexaploid oat evolution (Leggett, 1992). Haploid plants of hexaploid oat are of interest not only for producing doubled haploids as a breeding tool, but also as a genetic and cytogenetic tool for analyzing homoeologous chromosome pairing relationships (Nishiyama & Tabata, 1964) and for producing aneuploids (e.g., $2n - 1$ monosomics) for genetic mapping (Nishiyama *et al.*, 1968; Jellen *et al.*, 1993).

Only five haploid ($2n = 3x = 21$) or aneuhaploid oat plants of spontaneous origin have been reported in the literature. These were recovered from either twin embryo seeds or as progeny of crosses involving aneuploids (Leggett, 1977). One of these haploids proved particularly valuable as it was used to recover nine monosomic lines of oat as the basis of a monosomic series in the cultivar Kanota (Nishiyama *et al.*, 1968).

Two major approaches have been employed to produce haploid plants in cereals at frequencies high enough to be used as a breeding tool. These include culture of microspores either within anthers or isolated from them, and wide-hybrid crosses with rescue of developing embryos following chromosomal elimination of one of the parental genomes during early embryonic divisions. The amount of published literature on haploid production in oat is quite limited, partly because of the relatively few laboratories working on oat and partly because of a general lack of success in haploid production by anther or microspore culture in oat. Because of this limitation, this chapter focuses initially on our experiences at St. Paul, Minnesota, in anther culture in oat, briefly summarizes other oat anther/microspore culture efforts, and then is followed by descriptions of our recent discoveries on oat haploid plant production from wide hybridizations. Wide-cross hybrids of oat × maize (*Zea mays* L.) have produced oat haploid plants with some unique characteristics not found in haploids produced in other species. These include partial self-fertility in oat haploids (up to 50% seed set) yielding both euploid ($2n = 42$) and aneuploid progeny, and the discovery of plants recovered from oat × maize crosses that have retained one or more chromosomes of the pollen donor plant. These latter partial hybrids between members of different

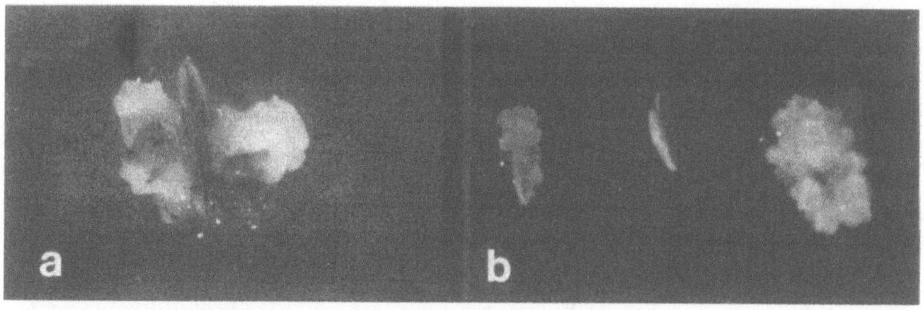


*Figure 1.* Cultured anthers of hexaploid oat cv. Stout with organized embryonic structures initiated (a) and with masses of unorganized callus initiated (b).

subfamilies of Gramineae (Pooideae and Panacoideae) represent probably the widest partial hybrids yet reported in which whole added alien chromosomes are present in stable, fertile lines.

## 2. Anther and microspore culture in oat

### 2.1. *Initial plant recovery from cultured oat anthers*

The first published study on anther culture in oat was by Chung (1980), who reported initiation of callus from oat anthers. She also obtained evidence from serial sections of callusing anthers for a microspore origin of these calli; however, no plants were recovered. The first haploid oat plant recovered via anther culture was reported from our group (Rines, 1983); however, only a single haploid (21 chromosome) plant and two euploid (42 chromosome) plants were recovered from about 65,000 anthers placed on various culture media, and these came from the same anther (Fig. 1a) although calli (Fig. 1b) had been initiated from 2,627 of the anthers. The cultivar Stout was the source of the anther giving the plants. It together with a related cultivar Clintford and their derivatives were the sources of most of the calli (up to 24% anthers callusing) among several tested genotypes. This strong genotypic influence on anther culture response has been observed throughout the cereals (various chapters, this book).

Growth conditions of the donor plants, heat and cold pre- and post-treatments of anthers, and media components including basic mineral salt levels, carbon and nitrogen sources and concentrations, liquid versus solidified media, and various growth regulator compounds and other additives all have been found to influence anther culture response markedly in various species. In oat, an MS (Murashige & Skoog, 1962) basal medium with 10% sucrose and no growth regulators gave the highest anther callus initiation frequency among media tested, but the only anther to produce plants had initially been plated on a modified potato extract medium containing 2 mg $L^{-1}$ 2,4-dichlorophenoxyacetic acid (2,4–D) and 0.5 mg $L^{-1}$ kinetin (Rines, 1983). This anther also was from a set that had been heat-shocked at 35°C for 24 h immediately after plating and prior to standard incubation at 22°C.

Oat anther callus initiated on agar-solidified MS media with 60 to 120 g $L^{-1}$ sucrose and no growth regulators tended to be highly unorganized cell masses (Rines, 1983). Most calli ceased growth after attaining a size of 1 to 5 $mm^3$; however, about 5 to 10%, depending on genotype, continued to grow as unorganized callus resulting in large masses which could be maintained in an active growing state for at least several months by periodic subculture on growth regulator-free medium. Such callus was often highly friable and formed well-dispersed suspension cultures that could be readily maintained on growth regulator-free MS liquid medium. One culture line (Stout 229–5) has been maintained for over ten years and has been distributed to various labs for use in metabolic and protoplast uptake and gene expression studies (e.g., Higgs & Colbert, 1993). Once this type of undifferentiated callus had initiated, however, we found no treatments including reduced sucrose, temperature shocks, addition of kinetin (0.2 to 2 mg $L^{-1}$), abscisic acid (0.01 to 1 mg $L^{-1}$), gibberellic acid (0.1 to 1 mg $L^{-1}$), or the purported antiauxin 2,4,6–trichlorophenoxyacetic acid (0.02 to 2 mg $L^{-1}$) that would elicit differentiated growth. Also, none of the compounds or treatments promoted either higher callus initiation or the development of tissue organization when they were included in the initial anther plating medium.

For several years following our initial recovery of a haploid oat plant from anther culture, we continued to excise and plate oat anthers involving a wide variety of treatments and procedures. New treatments included use of alternative carbon sources such as maltose, glucose, fructose, cellobiose, and oat, wheat, barley, and potato starches and also various thickening or gelling agents including Ficoll and starches. Among an additional over 100,000 anthers plated, we failed to recover another haploid oat plant, although on occasion organized tissues that appeared to have embryogenic potential were observed. More recently our efforts have shifted to wide hybrid crosses as a source of haploid oat plants, as will be described later.

### 2.2. *Oat anther/microspore culture*

Some successes in oat anther/microspore culture have been reported more recently from other laboratories. Polsoni (1991) at the University of Guelph reported consistent induction of microspore embryogenesis in oat anther culture. Tests of 15 cultivars and over 150 media modifications identified Stout as the highest responding cultivar and led to the development of a medium termed OAC-1. Nitrogen source and concentration were important, with the best embryo production occurring with glutamine at 700 mg $L^{-1}$ and ammonium nitrate at 250 mg $L^{-1}$. Maltose at 63 g $L^{-1}$ served as carbon source and potato-derived soluble starch at 60 g $L^{-1}$ as the gelling agent in the final medium. No growth regulators were present in the medium. An anther density of 10 anthers per mL medium and an incubation temperature of 25 to 30°C were most effective among treatments evaluated. Although numerous embryos were initiated from cultured oat anthers, no plants were obtained in this study.

Successful plant recovery from oat anther callus was reported by Sun *et al.* (1991), who recovered twelve green plants and one albino plant from anther culture of a naked (hulless) oat. In this report they found that 2,4-D in the medium was in general inhibitory to anther callus initiation. However, anthers placed onto MS medium with 1 mg $L^{-1}$ 2,4-D initiated a limited number of organized-type calli from which they were able to recover plants upon subsequent transfer to a regeneration medium of MS + 0.1 mg $L^{-1}$ naphthaleneacetic acid + 0.5 mg $L^{-1}$ benzyladenine. Root-tip cell chromosome counts on three of the green plants recovered revealed two were haploid and one was euploid. This success indicates that, at least in certain genotypes, production of haploid oat plants by anther culture is feasible, but to date only at a low frequency. Perhaps through the adaptation of new methods developed in other cereals or the introduction of some novel approach or identification of some key factor, anther or isolated microspore culture can be made sufficiently reliable to be used as a breeding tool in oat. Even then, anther culture as a means of haploid production in oat may be genotype-limited, just as it continues to be in the other cereals.

## 3. Oat haploids from oat × maize crosses

### 3.1. *Wide hybridization as an alternative source of oat haploids*

Based on a report by Laurie & Bennett (1986) of haploid wheat embryo formation resulting from wheat by maize crosses and our infrequent recovery of oat haploids from anther culture, we tried crossing oat × maize as a means to produce oat haploids. In our initial efforts 14 haploid oat plants were recovered from maize pollination of 3,300 emasculated oat florets (Rines & Dahleen, 1990). Since that time we have continued our oat wide-cross efforts with several modifications based on reports of success in wheat and other Triticeae of haploid production via wide crosses (see reviews by Laurie & O'Donoughue, 1994; Mujeeb-Kazi & Riera-Lizarazu, 1996). While there are many similarities in wide-cross production of haploids in wheat and oat, including chromosome elimination as the basic mechanism and a general lack of genotype restriction on either maternal or pollen donor parent, we have discovered several exciting unique features about oat haploid production by wide crosses that differ from that in wheat. These unique features include observations that a post-pollination auxin treatment is not essential for embryo development in oat wide crosses as it is in wheat, that oat haploids may be partially self-fertile via meiotic restitution, and, most significantly, that maize chromosome elimination is not always complete in oat wide crosses with maize. Many derived haploid oat plants retain one or more maize chromosomes, and these may be transmitted through sexual reproduction into subsequent generations. These novel characteristics of partial self-fertility and maize chromosome retention in oat haploids will be described later in this review.

### 3.2. *Factors affecting frequencies of embryo induction and haploid plant recovery*

The efficiency of haploid plant production via crosses to maize has remained much lower in oat than in wheat throughout our studies. We find, in general, 5 to 10% embryo initiation based on number of florets emasculated with oat versus our findings of about 25% in wheat and values in the literature commonly around 20 to 40% with as high as 56% (Laurie, 1989). Germination frequencies of rescued embryos also tend to be much lower in oat (5 to 15%) versus frequencies as high as 81% reported in wheat (Riera-Lizarazu & Mujeeb-Kazi, 1990). Thus, the overall frequency of haploid plants per pollinated floret is only 0.5 to 2% in oat vs. 10 to 30% in wheat. Also, oat florets are much more tedious to emasculate, pollinate, and post treat because of the physical arrangement of the oat spikelets (florets) in a panicle versus spikelets clustered in a spike in wheat. Furthermore, there is a spread in anthesis of several days for florets from top to bottom of the oat panicle, versus the more uniform maturation within a wheat spike. The low frequency of response plus the tediousness of manipulating florets has meant that it is much more difficult to conduct appropriate comparative experiments to quantify the effects of different factors on haploid oat plant recovery frequencies.

Even in wheat there appear to be several factors other than genotype which affect efficiency (Laurie, 1989; Laurie & O'Donoughue, 1994). Many of these are associated with the need to obtain plants of optimal vigor in both pollen donor and the recipient species that are synchronized in their flowering times, and the short window of time when each individual plant and floret is optimal for manipulation. For this reason we have found it advantageous to use greenhouse and growth chamber facilities, particularly for growing of our recipient maternal plants, with multiple planting dates of each. Even with this aid, it is difficult to obtain the synchrony and numbers needed for accurate comparisons. We have found that an alternative technique involving *in vitro* tiller culture, similar to that described by Riera-Lizarazu *et al.* (1992a) for wheat wide crosses, allows for use of field-grown plants, although we encounter more microbial embryo contamination problems in our rescue. Machan *et al.* (1995) have successfully used a tiller culture technique to recover oat haploids from crosses with maize using greenhouse grown oat plants.

The basic techniques used for growing plants, emasculating and pollinating florets, rescuing embryos, and cytologically analyzing rescued plants have been described (Rines & Dahleen, 1990). We have adapted significant modifications since then that include an auxin solution (usually 10 mg $L^{-1}$ 2,4-D) sprayed to saturation onto the cut florets 1–2 days post pollination and the use of 20 g $L^{-1}$ instead of 60 g $L^{-1}$ sucrose in the embryo rescue medium. Recent results (Maquieira & Rines, unpublished) indicate that a bilayer medium (Iglesias *et al.*, 1994) for rescuing young wheat embryos almost doubled the frequency of rescued oat embryos that germinated to green plants. In this technique embryos are placed onto blocks of high sucrose (150 g $L^{-1}$) medium that are then placed onto standard 30 g $L^{-1}$ sucrose medium. This arrangement provides a high initial osmoticum promoting embryo maturation followed by a decreasing osmoticum by diffusion to permit germination.

Evidence for fertilization followed by maize chromosome elimination as the origin of the haploid oat plants recovered from oat × maize crosses was provided from cytological observations of ovules 24 to 48 h post pollination in our initial report (Rines & Dahleen, 1990). Although no zygotic mitotic figures clearly showed chromosome complements of the two parents, as detailed by Laurie & Bennett (1986, 1988b, 1989) and Laurie *et al.* (1990) in crosses of wheat and other Triticeae with maize, pearl millet (*Pennisetum glaucum* [L.] R. Br.), and sorghum (*Sorghum bicolor* L.), the presence of chromosomes not incorporated in mitotic figures and numerous micronuclei in 28 to 36 h post-pollination endosperm divisions indicated that chromosome elimination had occurred in these products of oat × maize crosses. Conclusive evidence for a mechanism of fertilization and chromosome elimination came from haploid oat plants that retained one or more maize chromosomes (Riera-Lizarazu *et al.*, 1992b, 1996).

That fertilization of oat by maize is not restricted to specific genotypes of either species was indicated in our initial experiments where the first four haploid plants recovered were each from a different oat cultivar and each had been pollinated from a different maize inbred or hybrid (Rines & Dahleen, 1990). Possible genotypic effects on fertilization frequencies were further investigated in a study in which relative embryo and endosperm initiation frequencies were analyzed in 5-day post pollination ovules (Nunez, 1992). In one experiment, florets of seven oat genotypes were treated with pollen of a pearl millet and two maize genotypes, and in a second experiment florets of two oat genotypes were pollinated with a single pearl millet and eight maize genotypes. One to twelve embryos were found in all but seven of the 39 combinations, with about 100 emasculated florets pollinated for each combination. Data are shown in Table 1 for embryo only, endosperm only, and embryo plus endosperm initiation events detected at the 5-day stage for the two maternal oat genotypes and for the pollen donor pearl millet and two maize genotypes included in both experiments. Endosperm initiation, either alone or in combination with embryo initiation, occurred in all crosses. Laurie *et al.* (1990) reported large differences in relative frequencies of embryo and endosperm initiation in wide crosses involving the Triticeae with

*Table 1.* Embryo and endosperm formation detected five days post pollination in oat × maize and oat × pearl millet crosses[a]

| Pollen source | Oat parent | Experiment 1; Experiment 2[b] | | | | |
|---|---|---|---|---|---|---|
| | | No. ovules dissected | No. with embryos only | No. with endosperm only | No. with embryo and endosperm | Total % with embryos |
| A619XW64A (maize) | Starter | 93; 101 | 2; 2 | 2; 2 | 2; 3 | 4.3; 5.0 |
| | Lodi | 98; 183 | 1; 8 | 2; 3 | 0; 4 | 1.0; 14.4 |
| Honeycomb (maize) | Starter | 117; 100 | 3; 0 | 1; 1 | 0; 4 | 3.4; 4.0 |
| | Lodi | 107; 105 | 1; 2 | 1; 0 | 0; 9 | 1.0; 10.5 |
| Tift 383 (pearl millet) | Starter | 104; 93 | 2; 0 | 1; 4 | 4; 5 | 6.7; 5.4 |
| | Lodi | 101; 66 | 1; 1 | 0; 5 | 0; 4 | 1.0; 7.6 |

[a] Data from Nunez (1992).
[b] Data separated by a semi-colon are shown for two independent experiments.

the relative frequencies apparently being highly species specific. For example, endosperm initiation was infrequent in crosses of hexaploid wheat × maize whereas it was common in crosses of several tetraploid wheats × maize and also in hexaploid wheat × pearl millet crosses. The data in Table 1 also illustrate limitations on drawing conclusions regarding possible quantitative comparative differences among genotypes of either parent in oat × maize crosses. Overall frequencies tend to be low and undefined factors other than genotype appear to influence frequencies as illustrated by the large differences between Experiment 1 and Experiment 2 for crosses involving "Lodi" oat. We have noted that fertilization and embryo recovery frequencies tend to be reduced as a result of plant stress such as high temperature, drought, low light, and pesticide application, even when the parental plants remain highly self-fertile.

The further development of initiated embryos from oat × maize crosses appears to be less dependent on a post-pollination auxin treatment than does embryo development from wheat × maize crosses, although even in oat such auxin treatments may be beneficial. In wheat, only one haploid plant from 2,440 pollinated florets was obtained without a post pollination auxin treatment (Laurie & Bennett, 1988a). In oat × maize crosses, the initial 14 haploid oat plants recovered from 3,300 maize-pollinated florets were obtained without an auxin treatment (Rines & Dahleen, 1990). The effects of various levels of 2,4–D applied by spraying oat florets to saturation using a misting bottle at one day post pollination are illustrated in Table 2. Data are presented for frequencies of embryo only, endosperm only, embryo plus endosperm, and total embryos for caryopses dissected at 16 days post pollination.

*Table 2.* Effect of 2,4-D applied one day post pollination on frequency of embryo and endosperm presence in 16-day post pollination caryopses and on plant recovery from oat × maize crosses

| Oat genotype | 2,4-D (mg $L^{-1}$) | Panicles | Maize-pollinated florets[a] | Embryo only | Embryo plus endosperm | Endosperm only | % total with embryos | Total plants recovered |
|---|---|---|---|---|---|---|---|---|
| Starter-1 | 0 | 7 | 212 | 9 (1)[b] | 1 | 0 | 4.7 | (1) |
| | 10 | 8 | 282 | 12 (2) | 8 (1) | 3 | 7.1 | (3) |
| | 100 | 9 | 288 | 4 | 12 (1) | 16 | 5.6 | (1) |
| Sun II | 0 | 6 | 287 | 2 | 2 (1) | 0 | 1.4 | (1) |
| | 10 | 6 | 291 | 1 | 10 (4)[c] | 3 | 3.8 | (4)[c] |
| | 100 | 6 | 197 | 11 (4) | 0 | 8 | 5.6 | (4) |

[a] Pollen from "Seneca 60" sweetcorn was used throughout.

[b] The numbers of plants recovered from the various treatment combinations are given in parentheses.

[c] One of these four plants retained two maize chromosomes and one of these four retained four maize chromosomes. All other plants indicated in the table had only the haploid oat complement of 21 chromosomes.

Also shown are the numbers of plants recovered following plating of the excised embryos onto a rescue medium. Freshly collected pollen of "Seneca 60" maize was used with both oat genotypes. Although the scope of the experiment was limited, several observations can be made. Post-pollination treatments with 2,4-D appeared to be beneficial to embryo and plant recovery, and the effective range seemed to be broad. Based on this experiment and several others, we chose to use 10 mg $L^{-1}$ 2,4-D for post-pollination treatments for oat × maize crosses because the embryos appeared more normal in shape and with less browning of the caryopses than with higher 2,4-D concentrations. However, in a recent paper Machan *et al.* (1995) reported 24 haploid oat plants recovered from 34 embryos isolated from 751 maize-pollinated oat florets through the use of a 100 mg $L^{-1}$ 2,4-D treatment applied one day post pollination with application both by internode injection and placement of drops in each floret. Their post pollination treatment also involved spraying the panicles on the third, fifth, and seventh days with a solution consisting of 25 mg $L^{-1}$ gibberellic acid, 10 mg $L^{-1}$ benzyladenosine, 10 mg $L^{-1}$ beta-indoleacetic acid, and 15 mg $L^{-1}$ alpha-naphthaleneacetic acid.

Endosperm development in oat × maize caryopses in our experiments seemed to be enhanced by increased concentrations of 2,4-D in the post-pollination treatment (Table 2). The amount of endosperm, when detected in a caryopsis together with an embryo, varied greatly from barely detectable to an amount equal in size to the embryo. A "typical" dissected embryo plus endosperm from a 16-day post maize-pollinated oat caryopsis is shown in Fig. 2. The frequent detection of endosperm in 16-day post maize-pollinated oat caryopses differs from wheat × maize caryopses at this stage where de-

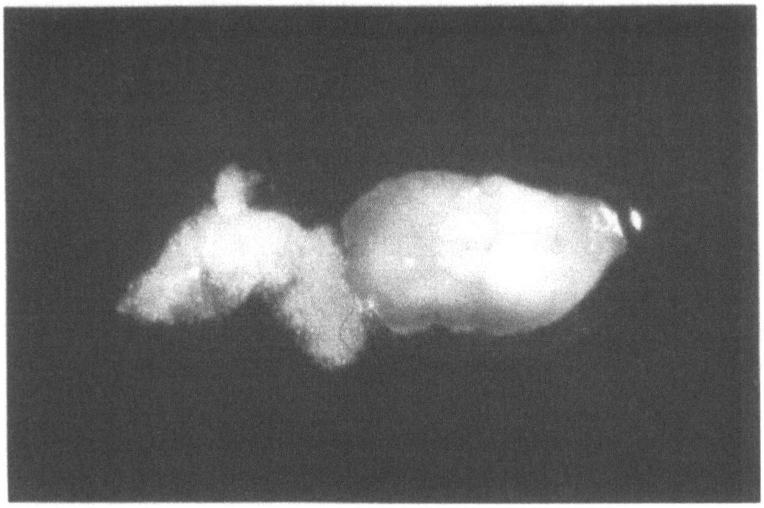

*Figure 2.* Embryo with poorly developed attached endosperm excised 16 days after pollination of oat with maize pollen.

tectable endosperm has rarely been present. The size of the embryos also varied greatly with a range from about 0.5 to 4 mm in length and with no apparent relation to either 2,4-D treatment or presence or absence of detectable endosperm.

Plant recovery from oat × maize crosses often is sporadic with no consistent association with endosperm being present or absent (Table 2). Plants usually originate from the mid to larger sized embryos but have been recovered from small (<1 mm) embryos as well. Overall plant recovery frequencies from oat × maize crosses in our experiments have been only about 0.5 to 2% of florets pollinated, but the recovered plants have some unique characteristics. Much of our recent research has focused on these plants.

### 3.3. *Partial self-fertility of oat haploids producing aneuploid and euploid progeny*

Haploid oat plants recovered from oat × maize crosses were often found to be partially self-fertile with seed set up to 50% in the primary and secondary florets of the main tillers (Rines & Dahleen, 1990). That this self-fertility was not due simply to spontaneous somatic doubling was indicated by the observation that partial seed set was scattered throughout a panicle, rather than sectored, and also many of the progeny were aneuploid with unexpected chromosome numbers (e.g., 39, 41, 43). Davis (1992) identified meiotic restitution during gamete formation as the probable source of this partial fertility. Analysis of microsporogenesis showed that functional gametes often arose through failure of the reduction division, thus producing 21-chromosome dyads rather than tetrads as meiotic products in the haploids. Monosomics (2n = 41) and other aneuploids were present at frequencies up to 25% of progeny from these self-fertile haploids. The high partial self-fertility discovered in the haploid oat plants from oat × maize crosses is probably not unique to oat haploids of wide-cross origin; rather, it may be that the number of oat haploid plants previously available was so few that this feature was not detected. Also, the partial self-fertility in haploid oat plants appears to be highest in cool, non-stress growing conditions such as in the growth chambers we use for plant rescue.

### 3.4. *Use of aneuploids derived from partially self-fertile oat haploids in genetic mapping*

Oat aneuploids recovered among progeny of partially self-fertile oat haploids derived from oat × maize crosses are currently being characterized for developing a complete monosomic series to use in mapping of molecular markers. We are participating in a large multi-institutional effort to develop a molecular marker linkage map for hexaploid oat. The recently published map is composed of 561 markers forming 38 linkage groups with 29 unlinked markers (O'Donoughue *et al.*, 1995). Association of markers to physical chromosomes through the use of aneuploids should allow condensing the map to the expected 21 linkage groups, one for each chromosome of the hexaploid oat genome complement ($n = 3x = 21$). Initial efforts employing cytological C-banding of root-tip cells and restriction fragment length polymorphisms (RFLP) revealed that the then available sets of oat monosomic stocks were incomplete, with a set in the cultivar Kanota having only 12 unique monosomics (Jellen *et al.*, 1993) and with three more unique monosomics contributed by a "SunII" set (unpublished results). Preliminary screening by chromosomal C-banding of monosomics recovered among progeny of partially self-fertile haploids derived from oat × maize crosses indicates that a complete set of oat monosomics may soon be available (E. Jellen, personal communication). Their identities are being confirmed using RFLP markers from the published marker linkage map. Thus, an integrated chromosomal/molecular marker linkage map may soon be available for oat.

### 3.5. *Haploids as a source of doubled haploids for oat breeding*

Although the presence of numerous aneuploids along with euploids in progeny of self-fertile haploids provides interesting cytogenetic variants for study, it also introduces complexities in trying to capitalize on the haploid self-fertile progeny as a source of "doubled haploids" for breeding purposes. Davis (1992) screened progeny from self-fertile haploids for low frequency of micronuclei (<5%) at the tetrad stage of microspore development as a way to select putatively normal 42-chromosome euploid progeny. In replicated field microplot tests of doubled haploid lines derived from cytologically screened progeny, 14 of 22 lines were equivalent to the parent "Starter" in yield, with the remainder yielding less. That the reduced performance of these lines was due to some chromosomal abnormality arising from the haploid self-fertility cannot be inferred at this time. Suenaga & Nakajima (1993) reported reduced performance in the majority of wheat lines derived from colchicine doubling of haploids from wheat × maize crosses. They postulated that much of the detrimental effect arose from the colchicine doubling. Colchicine was not used in our oat experiments.

### 3.6. *Retention of maize chromosomes in haploids from oat × maize crosses*

Probably the most significant difference between haploid plant production from oat × maize crosses and that from wide crosses involving wheat and other Triticeae members is that haploid oat plants often retain one or more maize chromosomes (Riera-Lizarazu *et al.*, 1992b, 1996). In root-tip cytological analysis of 90 plants recovered from over 15,000 maize-pollinated emasculated oat florets through seven sets of experiments, 30 had retained from one up to four maize chromosomes in addition to a full complement of 21 oat chromosomes (Fig. 3). That the smaller added chromosomes are indeed of maize origin was clearly shown using *in situ* genomic hybridization (Riera-Lizarazu *et al.*, 1992b, 1996). In the experiment shown in Table 2, two of the 14 recovered plants had maize chromosomes in their root-tip cells; one plant had four maize chromosomes and one had two. The other twelve had only a 21-chromosome oat complement. In addition to the 14 plants indicated in Table 2, four plantlets were initiated but lacked the vigor to survive transplanting to soil and no root tips were obtained for analysis. Plants that retained maize chromosomes, particularly those with multiple maize chromosomes, tended to be less vigorous and many died before flowering or were sterile. Thus, it is possible that these four less vigorous plantlets also carried added maize chromosomes.

### 3.7. *Transmission, identity, and effect of maize chromosome additions in oat*

Haploid oat plants with an individual maize chromosome have flowered and, through self-fertility by meiotic restitution, produced progeny retaining maize chromosomes as maize disomic additions ($2n = 42 + 2$) (Riera-Lizarazu *et al.*, 1996). Also, a haploid plant with two added maize chromosomes was self-fertile and produced a double disomic addition ($2n = 42 + 4$). Maize DNA sequences have been used as molecular marker probes to identify the maize chromosomes present as additions in these oat × maize derived lines (Rines *et al.*, 1995; Riera-Lizarazu *et al.*, 1996). A Southern hybridization

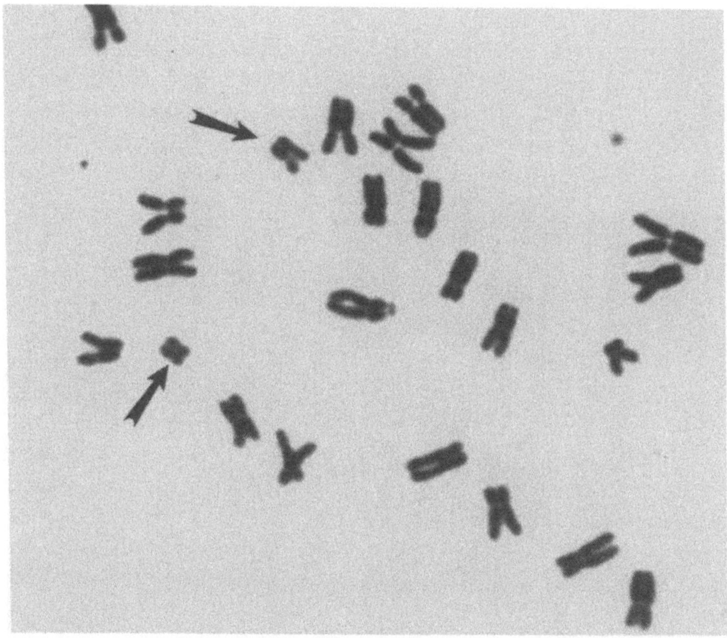

*Figure 3.* Root-tip cell of a plant recovered from oat × maize hybridization with 21 oat chromosomes and two maize chromosomes (arrows).

blot illustrating how the identification is made is shown in Fig. 4. It also may be noted that such blots made with DNA from a series of oat lines with single maize chromosome additions can be valuable in locating to a maize chromosome any DNA sequence probe that is either maize-specific or has a size pattern different from oat. Using these types of blots with our oat-maize additions, four disomic addition oat lines were identified as carrying maize chromosome 4, two with maize chromosome 9, and one each for maize chromosomes 2, 3, and 7. Maize chromosomes 4 and 7 were present in the double disomic addition line. Also, there was a monosomic addition line with maize chromosome 8. The non-fertile haploids included lines with added maize chromosomes 5 and 6. No plants retaining maize chromosomes 1 or 10 were recovered. One 21-chromosome plant had a segment of maize chromosome translocated onto an oat chromosome, as evidenced by cytological *in situ* hybridization analysis, but it was not fertile. We have not observed other maize-oat translocations in rather limited analyses conducted to date.

The oat-maize chromosome additions are the widest-cross stable, fertile partial hybrids of which we are aware. There are two published instances of possible alien chromosome retention in widely distant cereal crosses, one involving wheat and maize (Comeau *et al.*, 1992) and one in wheat and

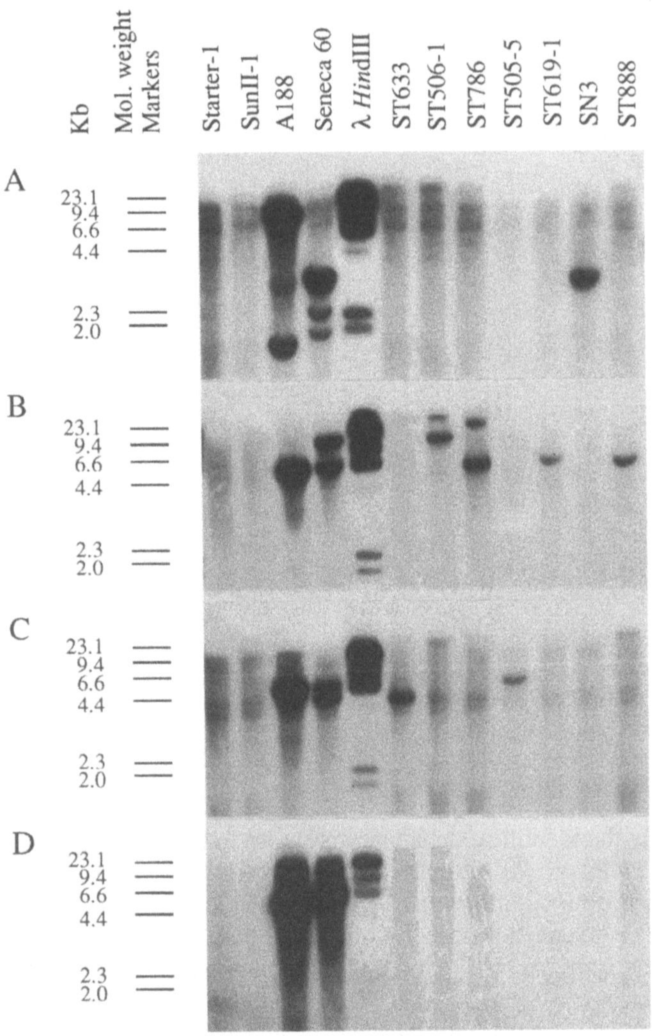


*Figure 4.* Maize chromosome identification in oat-maize derivatives using maize RFLP markers. (A) Southern blot hybridization of leaf DNA of oat (Starter-1 and SunII-1), maize (A188 and Seneca 60), and seven oat-maize derivatives (ST633, ST506-1, ST786, ST505-5, ST6191, SN3, and ST888) using umc16, a maize-chromosome 3 marker. Note the strong hybridization of the probe to DNA from maize and to a lesser degree to oat. One oat-maize derivative (SN3) has a band that corresponds to a band from its maize paternal parent Seneca 60. The other oat-maize derivatives lacked strong hybridizing bands. It is concluded that SN3 contained maize-chromosome 3. (B) Southern blot phybridization of the same individuals as in (A) using umc15, a maize-chromosome 4 marker. Note hybridization of the probe to maize but not to oat DNA. Four oat-maize derivatives (ST786, ST506-1, ST619-1, and ST888) have bands that correspond to bands in their maize parents. The other oat-maize lines lacked bands. It is concluded that ST786, ST506-1, ST619-1, and ST888 contained maize-chromosome 4. Also, ST506-1 and ST786 inherited different alleles of umc15 from Seneca 60, an $F_1$ hybrid line. (C) Southern blot

pearl millet (Ahmad & Comeau, 1989), but in neither case was the extra chromosome found to be transmitted to progeny.

In our oat-maize partial hybrids, the phenotypes compared to the oat maternal parent varied from little or no visible differences in plant appearance in the maize chromosome 4 and 9 additions to dramatic effects of liguleless upper leaf sheaths, crooked panicles, and adventitious shoots initiated from lower internodes in the maize chromosome 3 addition line. Even with these phenotypic alterations, the disomic chromosome 3 addition line was still highly fertile. Gene expression and frequencies of maize chromosome transmission are now being analyzed on these novel inter-subfamilial partial hybrids derived from oat × maize crosses.

## 4. Conclusions

Haploid plant recovery by anther or microspore culture in cultivated hexaploid oat has been infrequent with only two reports of successful plant recovery. Callus initiation has been more frequently observed, particularly with certain genotypes, but the callus is usually non-regenerable. The occurrence of spontaneous haploids from oat species crosses has also been rare, and the authors are unaware of any reports of haploid oat production via ovary culture or irradiated pollen – two methods that have been used in other species to produce haploids. However, wide-cross oat × maize hybridizations involving maize chromosome elimination during early embryonic cell divisions with subsequent embryo rescue has provided an alternative source of oat haploids. The wide hybridization procedure also appears to be much less restricted by genotype than anther culture. Haploid oat plants have the unique property of partial self-fertility due to meiotic restitution. This self-fertility yields both euploid progeny, which may serve as doubled haploid lines in a breeding program, and aneuploid progeny, which are being used to help in developing an integrated chromosomal/molecular linkage map in oat. Maize chromosome elimination is not always complete in oat × maize hybrid embryo development. Haploid oat plants retaining up to four maize

hybridization of the same individuals as in (A) using umc109, a maize-chromosome 9 marker. Note the strong hybridization of the probe to DNA from maize and to a lesser degree to oat. Two oat-maize derivatives (ST633 and ST505-1) have bands that correspond to bands from their maize paternal parent Seneca 60. The other oat-maize derivatives lacked hybridizing bands. It is concluded that ST633 and ST505–5 contained maize-chromosome 9. Also ST633 and ST505-1 inherited different alleles of umc109 from Seneca 60, an $F_1$ hybrid line. (D) Southern blot hybridization of the same individuals as in (A) using umc48, a maize-chromosome 8 marker. Note the strong hybridization of the probe to DNA from maize and no hybridization to oat or oat-maize derivatives. It is concluded that maize-chromosome 8 was absent in these oat-maize lines.

chromosomes have been identified. Ten plants with one and one plant with two maize chromosomes were self-fertile. With maize molecular marker probes, disomic maize chromosome-addition lines of oat were identified for six different maize chromosomes. A double disomic maize chromosome addition oat line was also identified that carried two maize chromosome pairs. The novel properties of partial self-fertility in haploid oat and the retention of maize chromosomes in fertile derivatives of oat × maize crosses provide exciting materials for gene transfer and mapping analyses.

## 5. Acknowledgements

Research support provided to the authors by The Quaker Oats Company and the Midwest Plant Biotechnology Consortium (USDA prime/Purdue Univ. sub #593–0120–13) is gratefully acknowledged.

## 6. References

Ahmad, F. & A. Comeau (1989) Wheat × pearl millet hybridization: consequences and potential. Euphytica 50: 181–190.

Chung, S.W. (1980) Anther culture in oats (*Avena sativa* L.). Ph.D. Thesis, McGill University, Montreal. Diss. Abstr. 41: 2845–B.

Comeau, A., P. Nadeau, A. Plourde, R. Simard, O. Maes, S. Kelly, L. Harper, J. Lettre, B. Landry & C.A. Pierre (1992) Media for in ovulo culture of proembryos of wheat and wheat-derived interspecific hybrids or haploids. Plant Sci. 81: 117–125.

Davis, D.W. (1992) Characterization of oat haploids and their progeny. M.S. Thesis, University of Minnesota, St. Paul.

Higgs, D.C. & J.T. Colbert (1993) $\beta$-glucuronidase gene expression and mRNA stability in oat protoplasts. Plant Cell Rep. 12: 445–452.

Iglesias, V.A., A. Gisel, I. Potrykus & C. Sautter (1994) *In vitro* germination of wheat proembryos to fertile plants. Plant Cell Rep. 13: 377–380.

Jellen, E.N., W.L. Rooney, R.L. Phillips & H.W. Rines (1993) Characterization of the hexaploid oat *Avena byzantina* cv. Kanota monosomic series using C-banding and RFLPs. Genome 36: 962–970.

Laurie, D.A. (1989) Factors affecting fertilization frequency in crosses of *Triticum aestivum* cv. Highbury × *Zea mays* cv. Seneca 60. Plant Breed. 103: 133–140.

Laurie, D.A. & M.D. Bennett (1986) Wheat × maize hybridization. Can. J. Genet. Cytol. 28: 313-316.

Laurie, D.A. & M.D. Bennett (1988a) The production of haploid wheat plants from wheat × - maize crosses. Theor. Appl. Genet. 76: 393–397.

Laurie, D.A. & M.D. Bennett (1988b) Chromosome behaviour in wheat × maize, wheat × - sorghum and barley × maize crosses. In: P.E. Brandham (Ed.), Kew Chromosome Conf. III, pp. 167–177. Her Majesty's Stationery Office, London.

Laurie, D.A. & M.D. Bennett (1989) The timing of chromosome elimination in hexaploid wheat × maize crosses. Genome 32: 953–961.

Laurie, D.A. & L.S. O'Donoughue (1994) Wheat × maize crosses for the production of wheat haploids. In: Y.P.S. Bajaj (Ed.), Maize. Biotechnology in Agriculture and Forestry, Vol. 25, pp. 102–118. Springer-Verlag, Berlin.

Laurie, D.A., L.S. O'Donoughue & M.D. Bennett (1990) Wheat × maize and other sexual hybrids: their potential for genetic manipulation and crop improvement. In: J.P. Gustafson (Ed.), Gene Manipulation in Plant Improvement II, pp. 95–126. Plenum Press, New York.

Leggett, J.M. (1977) The meiotic behaviour of aneuhaploids of the cultivated oat *Avena sativa* ($2n = 6x = 42$). Can. J. Genet. Cytol. 19: 651–656.

Leggett, J.M. (1992) Classification and speciation in *Avena*. In: H.G. Marshall & M.E. Sorrells (Eds.), Oat Science and Technology, pp. 29–52. Amer. Soc. Agron., Madison, WI.

Machan, F., Z. Nesvadba & L. Ohnoutkova (1995) Genetic stabilization and homogenization of new wheat and oat donors using haploidization with the aid of wide crossing. Genetika a Slecht. 31: 1–10.

Mujeeb-Kazi, A. & O. Riera-Lizarazu (1995) Polyhaploid production in the Triticeae by sexual hybridization. In: S.M. Jain, S.K. Sopory, & R.E. Veilleux (Eds.), *In Vitro* Haploid Production in Higher Plants, Vol. 1, Kluwer Academic Publishers, Dordrecht, pp. 275–296.

Murashige, T. & F. Skoog (1962) A revised medium for rapid growth and bioassays with tobacco tissue cultures. Physiol. Plant. 15: 473–497.

Nishiyama, I., R.A. Forsberg, H.L. Shands & M. Tabata (1968) Monosomics of Kanota oats. Can. J. Genet. Cytol. 10: 601–612.

Nishiyama, I. & M. Tabata (1964) Cytogenetic studies in *Avena*. XII. Meiotic chromosome behavior in a haploid cultivated oat. Jpn. J. Genet. 38: 311–316.

Nunez, V.M. (1992) Embryo and endosperm initiation in intergeneric crosses with oat (*Avena sativa* L.). M.S. Thesis, University of Minnesota, St. Paul.

O'Donoughue, L.S., S.F. Kiamian, P.J. Rayapati, G.A. Penner, M.E. Sorrells, S.D. Tanksley, R.L. Phillips, H.W. Rines, M. Lee, G. Fedak, S.J. Molnar, D. Hoffman, C.A. Salas, B. Wu, E. Autrique & A. VanDeynze (1995) A molecular linkage map of cultivated oat. Genome 38: 368–380.

Polsoni, L. (1991) The induction of microspore embryogenesis in anther culture of oats (*Avena sativa* L.). M.S. Thesis, University of Guelph, Guelph.

Riera-Lizarazu, O. & A. Mujeeb-Kazi (1990) Maize (*Zea mays* L.) mediated wheat (*Triticum aestivum* L.) polyhaploid production using various crossing methods. Cereal Res. Commun. 18: 339–345.

Riera-Lizarazu, O., A. Mujeeb-Kazi & M.D.H.M. William (1992a) Maize (*Zea mays* L.) mediated polyhaploid production in some Triticeae using a detached tiller method. J. Genet. Breed. 46: 335–346.

Riera-Lizarazu, O., H.W. Rines & R.L. Phillips (1992b) Retention of maize chromosomes in haploid oat plants from oat × maize crosses. Amer. Soc. Agron. Abstr.: 112.

Riera-Lizarazu, O., H.W. Rines & R.L. Phillips (1996) Cytological and molecular characterization of oat × maize partial hybrids. Theor. Appl. Genet. 92: (in press).

Rines, H.W. (1983) Oat anther culture: Genotype effects on callus initiation and the production of a haploid plant. Crop Sci. 23: 268–272.

Rines, H.W. & L.S. Dahleen (1990) Haploid oat plants produced by application of maize pollen to emasculated oat florets. Crop Sci. 30: 1073–1078.

Rines, H.W., O. Riera-Lizarazu & R.L. Phillips (1995) Disomic maize chromosome-addition oat plants derived from oat × maize crosses. K. Oono & F. Takaiwa (Eds.), Modification of Gene Expression and Non-Mendelian Inheritance, pp. 235–251. Natl. Inst. Agrobiol. Resources, Tsukuba, Japan.

Suenaga, K. & K. Nakajima (1993) Variation in doubled haploid plants of wheat obtained through wheat (*Triticum aestivum*) × maize (*Zea mays*) crosses. Plant Breed. 111: 120–124.

Sun, C.S., T.G. Lu & M.R. Sondahl (1991) Anther culture of naked oat and the establishment of its haploid suspension cell. Acta Bot. Sin. 33: 417–420.

*In Vitro* Haploid Production in Higher Plants
*Volume 4* – Cereals

S.B. Andersen, The Royal Veterinary and Agricultural University, Plant Breeding and Biotechnology, 40 Thorvaldsensvej, DK-1871, Fredriksberg C, Copenhagen, Denmark. Fax: +46 35 28 34 68.

B. Bohanec, Center for Plant Biotechnology and Breeding, Agronomy Department, Biotechnical Faculty, Jamnikarjeva 101, Ljubljana 61 111, Slovenia. Fax: +38 61 261 073.

B. Büter, Swiss Federal Institute of Technology (ETH) Zurich, Institute of Plant Sciences, Division of Agronomy and Plant Breeding, ETH-Eschikon 33, CH-8315 Lindau, Switzerland. Fax: ++41 52 33 27 06.

B.-H. Choi, Crop Experiment Station, RDA, Suwon 441–100, Korea. Fax: ++82 331 292 4560.

D.W. Davis, Department of Agronomy and Plant Genetics and Plant Molecular Genetics Institute, University of Minnesota, St. Paul, MN 55108, USA.

S. Deimling, Universität Hohenheim, Forschungsschwerpunkt Biotechnologie und Pflanzenzuchtung, P.O. Box 70 05 62 (350), 7000 Stuttgart 70, Germany. Fax:+49 711 459 2343.

H.S. Dhaliwal, Biotechnology Center, Punjab Agricultural University, Ludhiana, 141 004, India.

T. Flehinghaus-Roux, University of Hohenheim, Forschungsschwerpunkt Biotechnologie und Pflanzenzuchtung, P.O. Box 70 05 62 (350), 7000 Stuttgart 70, Germany.

B.P. Forster, Cell and Molecular Genetics Department, Scottish Crop Research Institute, Invergowrie, Dundee DD2 5D, Scotland.

R. Sandhu-Gill, Biotechnology Center, Punjab Agricultural University, Ludhiana, 141 004, India.

S.S. Gosal, Center of Biotechnology, Punjab Agricultural University, Ludhiana, India. Fax: ++91 161 400 945.

X. Gu, Department of Agronomy and Fort Hays Experiment Station, Kansas State University, Manhattan, KS 66506-5501, USA.

N. Halberg, Royal Veterinary and Agricultural University, Department of Agricultural Science, Thorvaldsensvej 40, DK 1871, Frederiksberg C, Denmark.

H. Hu, Institute of Genetics, Academia Sinica, Beijing, China.

G.S. Khehra, Biotechnology Center, Punjab Agricultural University, Ludhiana, 141 004, India.

K.D. Kofoid, Department of Agronomy and Fort Hays Experiment Station, Kansas State University, Manhattan, KS 66506-5501, USA.

G.H. Liang, Department of Agronomy, Kansas State University, Throckmorton Hall, Manhattan, KS 66506, USA. Fax: 913-532-6094.

S. Madsen, Royal Veterinary and Agricultural University, Department of Agricultural Science, Thorvaldsensvej 40, DK 1871, Frederiksberg C, Denmark.

V.M. Nunez, Department of Agronomy and Plant Genetics and Plant Molecular Genetics Institute, University of Minnesota, St. Paul, MN 55108, USA.

A. Olesen, Royal Veterinary and Agricultural University, Department of Agricultural Science, Thorvaldsensvej 40, DK 1871, Frederiksberg C, Denmark.

K.-Y. Park, Crop Experiment Station, RDA, Suwon 441-100, Korea.

R.-K. Park, Crop Experiment Station, RDA, Suwon, 441-100, Korea.

R.L. Phillips, Department of Agronomy and Plant Genetics and Plant Molecular Genetics Institute, University of Minnesota, St. Paul, MN 55108, USA.

W. Powell, Scottish Crop research Institute, Invergowrie, Dundee DD2 5DA, UK. Fax: 382-562 426.

O. Riera-Lizarazu, Department of Agronomy and Plant Genetics and Plant Molecular Genetics Institute, University of Minnesota, St Paul, MN 55108, USA.

H.W. Rines, USDA, ARS, 411 Borlaug Hall, University of Minnesota, 1991 Upper Buford Circle, St. Paul, MN 55108, USA. Fax: 612-625-1268.

N. Roulund, DLF-TRIFOLIUM, Danish Plant Breeding A/S, Hojerupvej 31, DK 4660, Storè Heddinge, Denmark.

P. Ryöppy, Department of Applied Chemistry and Microbiology, University of Helsinki, SF-00014, Box 27, Helsinki, Finland. Fax: 358-0-708 5475.

J.S. Sandhu, Biotechnology Center, Punjab Agricultural University, Ludhiana, 141 004, India.

G.S. Sidhu, Regional Rice Research Station, Punjab Agricultural University, Ludhiana, 141 004, India.

A.S. Sindhu, Biotechnology Center, Punjab Agricultural University, Ludhiana, 141 004, India.

B. Singh, Biotechnology Center, Punjab Agricultural University, Ludhiana, 141 004, India.

I.K. Vasil, Laboratory of Plant Cell and Molecular Biology, Department of Horticultural Sciences, 1143 Fifield Hall, University of Florida, Gainesville, FL 32611-0690, USA. Fax: 352 392 9366.

G. Yue, Department of Agronomy and Fort Hays Experiment Station, Kansas State University, Manhattan, KS 66506-5501, USA.

# Species and subject index

## Current Plant Science and Biotechnology in Agriculture

1. H.J. Evans, P.J. Bottomley and W.E. Newton (eds.): *Nitrogen Fixation Research Progress.* Proceedings of the 6th International Symposium on Nitrogen Fixation (Corvallis, Oregon, 1985). 1985 ISBN 90-247-3255-7
2. R.H. Zimmerman, R.J. Griesbach, F.A. Hammerschlag and R.H. Lawson (eds.): *Tissue Culture as a Plant Production System for Horticultural Crops.* Proceedings of a Conference (Beltsville, Maryland, 1985). 1986 ISBN 90-247-3378-2
3. D.P.S. Verma and N. Brisson (eds.): *Molecular Genetics of Plant-microbe Interactions.* Proceedings of the 3rd International Symposium on this subject (Montréal, Québec, 1986). 1987 ISBN 90-247-3426-6
4. E.L. Civerolo, A. Collmer, R.E. Davis and A.G. Gillaspie (eds.): *Plant Pathogenic Bacteria.* Proceedings of the 6th International Conference on this subject (College Park, Maryland, 1985). 1987 ISBN 90-247-3476-2
5. R.J. Summerfield (ed.): *World Crops: Cool Season Food Legumes.* A Global Perspective of the Problems and Prospects for Crop Improvement in Pea, Lentil, Faba Bean and Chickpea. Proceedings of the International Food Legume Research Conference (Spokane, Washington, 1986). 1988 ISBN 90-247-3641-2
6. P. Gepts (ed.): *Genetic Resources of* Phaseolus *Beans.* Their Maintenance, Domestication, Evolution, and Utilization. 1988 ISBN 90-247-3685-4
7. K.J. Puite, J.J.M. Dons, H.J. Huizing, A.J. Kool, M. Koorneef and F.A. Krens (eds.): *Progress in Plant Protoplast Research.* Proceedings of the 7th International Protoplast Symposium (Wageningen, The Netherlands, 1987). 1988 ISBN 90-247-3688-9
8. R.S. Sangwan and B.S. Sangwan-Norreel (eds.): *The Impact of Biotechnology in Agriculture.* Proceedings of the International Conference The Meeting Point between Fundamental and Applied in vitro Culture Research (Amiens, France, 1989). 1990. ISBN 0-7923-0741-0
9. H.J.J. Nijkamp, L.H.W. van der Plas and J. van Aartrijk (eds.): *Progress in Plant Cellular and Molecular Biology.* Proceedings of the 8th International Congress on Plant Tissue and Cell Culture (Amsterdam, The Netherlands, 1990). 1990 ISBN 0-7923-0873-5
10. H. Hennecke and D.P.S. Verma (eds.): *Advances in Molecular Genetics of Plant–Microbe Interactions.* Volume 1. 1991 ISBN 0-7923-1082-9
11. J. Harding, F. Singh and J.N.M. Mol (eds.): *Genetics and Breeding of Ornamental Species.* 1991 ISBN 0-7923-1094-2
12. J. Prakash and R.L.M. Pierik (eds.): *Horticulture – New Technologies and Applications.* Proceedings of the International Seminar on New Frontiers in Horticulture (Bangalore, India, 1990). 1991 ISBN 0-7923-1279-1
13. C.M. Karssen, L.C. van Loon and D. Vreugdenhil (eds.): *Progress in Plant Growth Regulation.* Proceedings of the 14th International Conference on Plant Growth Substances (Amsterdam, The Netherlands, 1991). 1992 ISBN 0-7923-1617-7
14. E.W. Nester and D.P.S. Verma (eds.): *Advances in Molecular Genetics of Plant–Microbe Interactions.* Volume 2. 1993 ISBN 0-7923-2045-X
15. C.B. You, Z.L. Chen and Y. Ding (eds.): *Biotechnology in Agriculture.* Proceedings of the First Asia-Pacific Conference on Agricultural Biotechnology (Beijing, China, 1992). 1993 ISBN 0-7923-2168-5

## Current Plant Science and Biotechnology in Agriculture

16. J.C. Pech, A. Latché and C. Balagué (eds.): *Cellular and Molecular Aspects of the Plant Hormone Ethylene*. 1993 ISBN 0-7923-2169-3
17. R. Palacios, J. Mora and W.E. Newton (eds.): *New Horizons in Nitrogen Fixation*. Proceedings of the 9th International Congress on Nitrogen Fixation (Cancún, Mexico, 1992). 1993 ISBN 0-7923-2207-X
18. Th. Jacobs and J.E. Parlevliet (eds.): *Durability of Disease Resistance*. 1993 ISBN 0-7923-2314-9
19. F.J. Muehlbauer and W.J. Kaiser (eds.): *Expanding the Production and Use of Cool Season Food Legumes*. A Global Perspective of Peristent Constraints and of Opportunities and Strategies for Further Increasing the Productivity and Use of Pea, Lentil, Faba Bean, Chickpea, and Grasspea in Different Farming Systems. Proceedings of the Second International Food Legume Research Conference (Cairo, Egypt, 1992). 1994 ISBN 0-7923-2535-4
20. T.A. Thorpe (ed.): In Vitro *Embryogenesis in Plants*. 1995 ISBN 0-7923-3149-4
21. M.J. Daniels, J.A. Downie and A.E. Osbourn (eds.): *Advances in Molecular Genetics of Plant-Microbe Interactions*. Volume 3. 1994 ISBN 0-7923-3207-5
22. M. Terzi, R. Cella and A. Falavigna (eds.): *Current Issues in Plant Molecular and Cellular Biology*. Proceedings of the VIIIth International Congress on Plant Tissue and Cell Culture (Florence, Italy, 1994). 1995 ISBN 0-7923-3322-5
23. S.M. Jain, S.K. Sopory and R.E. Veilleux (eds.): In Vitro *Haploid Production in Higher Plants*. Volume 1: Fundamental Aspects and Methods. 1996 ISBN 0-7923-3577-5
24. S.M. Jain, S.K. Sopory and R.E. Veilleux (eds.): In Vitro *Haploid Production in Higher Plants*. Volume 2: Applications. 1996 ISBN 0-7923-3578-3
25. S.M. Jain, S.K. Sopory and R.E. Veilleux (eds.): In Vitro *Haploid Production in Higher Plants*. Volume 3: Important Selected Plants. 1996 ISBN 0-7923-3579-1
26. S.M. Jain, S.K. Sopory and R.E. Veilleux (eds.): In Vitro *Haploid Production in Higher Plants*. Volume 4: Cereals. 1996 ISBN 0-7923-3978-9
27. I.A. Tikhonovich, N.A. Provorov, V.I. Romanov and W.E. Newton (eds.): *Nitrogen Fixation: Fundamentals and Applications*. 1995 ISBN 0-7923-3707-7
28. N.L. Taylor and K.H. Quesenberry: *Red Clover Science*. 1996 ISBN 0-7923-3887-1
29. S.M. Jain, S.K. Sopory and R.E. Veilleux (eds.): In Vitro *Haploid Production in Higher Plants*. Volume 5: Oil, Ornamental and Miscellaneous Plants. 1996 ISBN 0-7923-3979-7
30. R.H. Ellis, M. Black, A.J. Murdoch and T.D. Hong (eds.): *Basic and Applied Aspects of Seed Biology*. 1997 ISBN 0-7923-4363-8

KLUWER ACADEMIC PUBLISHERS – DORDRECHT / BOSTON / LONDON

MIX
Papier aus verantwortungsvollen Quellen
Paper from responsible sources
FSC® C105338

If you have any concerns about our products,
you can contact us on
**ProductSafety@springernature.com**

In case Publisher is established outside the EU,
the EU authorized representative is:
**Springer Nature Customer Service Center GmbH**
**Europaplatz 3, 69115 Heidelberg, Germany**

Printed by Libri Plureos GmbH
in Hamburg, Germany